CONTENTS

REVIEW IRAQI AGRICULTURE: FROM OIL FOR FOOD TO THE FUTURE OF IRAQI PRODUCTION, AGRICULTURE AND TRADE

HEARING

BEFORE THE

COMMITTEE ON AGRICULTURE
HOUSE OF REPRESENTATIVES

ONE HUNDRED EIGHTH CONGRESS

SECOND SESSION

JUNE 16, 2004

Serial No. 108–33

Printed for the use of the Committee on Agriculture
www.agriculture.house.gov

U.S. GOVERNMENT PRINTING OFFICE

94–706 PDF WASHINGTON : 2004

For sale by the Superintendent of Documents, U.S. Government Printing Office
Internet: bookstore.gpo.gov Phone: toll free (866) 512–1800; DC area (202) 512–1800
Fax: (202) 512–2250 Mail: Stop SSOP, Washington, DC 20402–0001

COMMITTEE ON AGRICULTURE

BOB GOODLATTE, Virginia, *Chairman*

JOHN A. BOEHNER, Ohio
Vice Chairman
RICHARD W. POMBO, California
NICK SMITH, Michigan
TERRY EVERETT, Alabama
FRANK D. LUCAS, Oklahoma
JERRY MORAN, Kansas
WILLIAM L. JENKINS, Tennessee
GIL GUTKNECHT, Minnesota
DOUG OSE, California
ROBIN HAYES, North Carolina
CHARLES W. "CHIP" PICKERING, Mississippi
TIMOTHY V. JOHNSON, Illinois
TOM OSBORNE, Nebraska
MIKE PENCE, Indiana
DENNIS R. REHBERG, Montana
SAM GRAVES, Missouri
ADAM H. PUTNAM, Florida
MAX BURNS, Georgia
JO BONNER, Alabama
MIKE ROGERS, Alabama
STEVE KING, Iowa
CHRIS CHOCOLA, Indiana
MARILYN N. MUSGRAVE, Colorado
DEVIN NUNES, California
RANDY NEUGEBAUER, Texas

CHARLES W. STENHOLM, Texas,
Ranking Minority Member
COLLIN C. PETERSON, Minnesota
CALVIN M. DOOLEY, California
TIM HOLDEN, Pennsylvania
BENNIE G. THOMPSON, Mississippi
MIKE McINTYRE, North Carolina
BOB ETHERIDGE, North Carolina
BARON P. HILL, Indiana
JOE BACA, California
RICK LARSEN, Washington
MIKE ROSS, Arkansas
ANÍBAL ACEVEDO-VILÁ, Puerto Rico
ED CASE, Hawaii
RODNEY ALEXANDER, Louisiana
DENNIS A. CARDOZA, California
DAVID SCOTT, Georgia
JIM MARSHALL, Georgia
BEN CHANDLER, Kentucky
STEPHANIE HERSETH, South Dakota
EARL POMEROY, North Dakota
LEONARD L. BOSWELL, Iowa
MARK UDALL, Colorado
RICK LARSEN, Washington
LINCOLN DAVIS, Tennessee

PROFESSIONAL STAFF

WILLIAM E. O'CONNER, JR., *Staff Director*
KEVIN KRAMP, *Chief Counsel*
STEPHEN HATERIUS, *Minority Staff Director*
ELYSE BAUER, *Communications Director*

REVIEW IRAQI AGRICULTURE: FROM OIL FOR FOOD TO THE FUTURE OF IRAQI PRODUCTION, AGRICULTURE AND TRADE

WEDNESDAY, JUNE 16, 2004

HOUSE OF REPRESENTATIVES,
COMMITTEE ON AGRICULTURE,
Washington, DC.

The committee met, pursuant to call, at 10:00 a.m., in room 1300 of the Longworth House Office Building, Hon. Bob Goodlatte (chairman of the committee) presiding.

Present: Representatives Smith, Moran, Ose, Osborne, Rehberg, Burns, Bonner, Rogers, Stenholm, Peterson, Dooley, Etheridge, Alexander, Scott, Herseth, Pomeroy, and Boswell.

Staff present: Brent Gattis, Lynn Gallagher, Matt O'Mara, Jason Vaillancourt, Callista Gingrich, clerk; Brandon Farris, and Andy Baker.

OPENING STATEMENT OF HON. BOB GOODLATTE, A REPRESENTATIVE IN CONGRESS FROM THE COMMONWEALTH OF VIRGINIA

The CHAIRMAN. Good morning. This hearing of the House Committee on Agriculture will come to order. At this time, I would like to recognize the gentleman from Texas, Mr. Stenholm, for the purpose of making an introduction.

Mr. STENHOLM. Thank you, Mr. Chairman. It is my distinct to welcome the newest Member to the House Agriculture Committee, Stephanie Herseth of South Dakota, just winning the special election. She comes to us from a farm background, has spent a considerable amount of her time dealing with rural issues in South Dakota prior to being elected to the Congress. So Stephanie, welcome. Glad to have you here.

The CHAIRMAN. Congresswoman Herseth, I want to add my congratulations to you, as well, and welcome to the House Agriculture Committee. We have a very congenial working relationship across the aisle, and we look forward to that kind of relationship with you as well.

On behalf of the committee, I welcome our distinguished witnesses. Today we will review agriculture in Iraq. This hearing will focus on the successful food assistance operation that enabled the U.S. and the World Food Program to feed 26 million Iraqis during the transition from Saddam's Iraq to a free Iraq. We will also receive an update on the investigations into the Oil for Food Program, and we will have a discussion about the future of Iraqi production, agriculture and trade.

For most of its history, Iraq has maintained a strong agricultural sector. The area of the Fertile Crescent in ancient Mesopotamia, which lies on a plain between the Tigris and Euphrates rivers, gave rise to the first known agricultural settlements 11,000 years ago. Unfortunately, Saddam Hussein was destined to forget the history of this cradle of civilization and did everything in his power to neglect the agricultural industry of his country. Iraqi agriculture research, improvements to infrastructure and advancements in technology were essentially nonexistent during Saddam's tenure and much of Iraqi agriculture is a product of the 1950's. Saddam also destroyed one of the world's largest wetland ecosystems when he drained the marshlands of southern Iraq, displacing thousands of Iraqi farmers. I am confident that the new leadership of free Iraq will not treat this important industry or the Iraqi farmers in the same manner in which Saddam's tyrannical regime did.

The United States has and will continue to work with the Iraqi government to make sure that the food security of the Iraqi people is maintained. One of the most overlooked successes of Operation Iraqi Freedom was that the humanitarian crisis that was predicted by many never materialized. Many predicted a humanitarian disaster and a mass exodus of refugees from Iraq in the spring and summer of 2003, but that disaster never occurred. This was because of the work of USAID and the World Food Program in reestablishing the Iraqi Public Distribution System immediately after the conflict. Today, you will hear how the U.S. Government worked closely with the World Food Program by donating over 2 million metric tons of food to the Iraqi people before, during and immediately after the conflict. Food security was maintained and 26 million Iraqis were fed.

Even though this food assistance to Iraq was provided temporarily during the transition last summer, the importance and effectiveness of the U.S. Food Aid Program in preventing disruptions to the food supply and keeping needy people fed is clear. Currently in the World Trade Organization negotiations, some countries, including the European Union, are criticizing the use of food aid. Opponents of U.S. food aid prefer that government-to-government "in kind" food aid be banned, and that a U.N. agency sanction all food aid, and that non-emergency food aid only be allowed to be given in the form of cash to purchase food from the world market.

Each year, 10 million people die from hunger and malnutrition. With millions starving around the world in Africa and Asia, I am deeply troubled by the proposals of these countries which would effectively end the U.S. Food Aid Program. The executive director of the World Food Program, Mr. James Morris, recently spoke of the advantages of food aid. Mr. Morris said, "Is more food aid the answer? Yes, definitely, but only part of it. We need more investment in agricultural infrastructure and scientific innovation, including biotechnology. Well-targeted food aid has tremendous potential, especially for reaching poor women and children. I would match food aid up against any kind of assistance on several important counts." I am committed to maintaining U.S. Food Aid Programs so that the bounty of American's farmers and ranchers can be shared with the needy around the world.

While the food security of the Iraqi people was a primary concern for the planners of Operation Iraqi Freedom and the Coalition Provisional Authority that followed, the food security of the Iraqi people did not seem to be a concern of Saddam Hussein. The fact that the Iraqi people were starving in the early 1990's did not seem to curtail the building of Saddam's elaborate palaces all across Iraq.

The United Nations' sanctions that were implemented after the first Gulf War in 1990 allowed for the trading of food and medicine, but it was evidence that Saddam's regime did not make feeding the Iraqi people a priority. The Saddam regime had twice denied the United Nations' proposals to establish an Oil for Food Program because he was not satisfied with the power he would have with the program. Once the Oil for Food Program did get off the ground in May 1996, Saddam had already devised his scheme to take advantage of the program. The United Nations allowed Saddam's regime to maintain sovereignty by negotiating contracts and deciding on the contractors for oil sales and the purchase the humanitarian goods.

Even though the United Nations did not have the final say in which Oil for Food contracts were approved, the kickbacks, smuggled oil and after-sales service charges reported by the General Accounting Office were enough for Saddam to pocket over $10 billion from the program. The Oil for Food scandal, which is currently being investigated, is of great concern to this committee and the American taxpayer. I am hopeful that the investigations currently underway in the Congress, in the United Nations, and in the Interim Iraqi Government will shed some light on ways in which the United Nations can learn from the lessons of the past and hold all those involved in this scandal accountable.

In the meantime, we must look to the future of Iraqi agriculture by helping Iraq rebuild its agricultural sector. The rebuilding has begun, and on May 5, 2004, the Iraqi people assumed full authority of the Ministry of Agriculture. I am confident that the Agriculture Ministry will work to transform Iraqi agriculture into a market-based sector.

Iraq can be self-sufficient in certain products like fruits and vegetables, but it is anticipated that they will have to import wheat, rice and feed grains. In 1989, the U.S. exported over $1.1 billion worth of agricultural products to Iraq. At that time, Iraq was our 12th largest customer. America's farmers and ranchers want to once again become a reliable supplier to the Iraqi people. The Coalition Provisional Authority and the Iraqi Ministry of Trade, with the help of USAID and USDA, have been working together to move Iraqi agriculture towards a more democratic and market-based system. I am hopeful that Iraq can make the transition to a market-driven economy and that America's farmers and ranchers will soon be making commercial sales to the Iraqi people.

I look forward to hearing from our witnesses today, and at this time, it is my pleasure to recognize the gentleman from Texas, Mr. Stenholm.

OPENING STATEMENT OF HON. CHARLES W. STENHOLM, A REPRESENTATIVE IN CONGRESS FROM THE STATE OF TEXAS

Mr. STENHOLM. Thank you, Mr. Chairman, and thank you for calling this hearing this morning to examine the U.N. Oil for Food Program and the transition to a more normal trading relationship between Iraq and its trading partners, which wee all hope will include American agriculture.

Iraq has traditionally been an important market for U.S. rice and wheat producers and has the potential to be so again, if we are willing to use our WTO Legal Export Program, such as the Export Credit Guaranty Program, the Market Access Program, and the Foreign Market Development Program. Between 1984 and 1989, Iraq imported over $2.2 billion in agricultural products each year, with a quarter of that coming from the United States. In the case of cereals, the U.S. dominated, with a 59 percent share of the Iraqi market. Almost all of those sales were made under U.S. Government programs. From 1983 to mid–1990, Iraq received nearly $5 billion in U.S. export credit guarantees, and by 1989, Iraq was our 12th largest agricultural export market.

Since the first food shipments began under the U.N. Oil for Food Program in March 1997, the Australian Wheat Board has dominated Iraq's wheat trade, with a 73 percent share, and the U.S. has been virtually excluded from the Iraq rice market, with Vietnam, Thailand and China each taking a share. As we will hear today from the General Accounting Office, the former Iraq regime acquired $4.4 billion through surcharges on oil sales and illicit commissions from suppliers exporting goods to Iraq through the Oil for Food Program, World Food Program has pointed out. Given this background, I think we ought to be looking closely at how soon we can get our export programs back up and running. In a recent response to a rice producer from Texas, Under Secretary Penn stated the following.

> Resumption of the GSM Program would require a multi-step process, involving forgiveness or rescheduling or both of all debt by the Paris Club, review by the Interagency Country Risk Assessment System, congressional appropriation of the funds needed to forgive any U.S. debt, and the assignment of credit lines to banks approved by the Commodity Credit Corporation. This can be a lengthy process.

Others have suggested that there is no regulatory or consistent administrative rules that require the Secretary of Agriculture to find that Iraq is not credit-worthy due to the $4 billion arrears owed by Iraq to the United States. In fact, they argue that the credit-worthiness determination should be forward looking, and that the issue should be whether a new regime in Iraq would have the means and inclination to repay new GSM debt in the future. I will not put Mr. Schatz on the spot by asking him to choose between his boss's view and the other view, but I would like to encourage USDA to continue its review of the appropriateness of using export credit guarantees in Iraq in the near future.

Another important aspect of the role of agriculture in Iraq's transition is food aid, and I would like to congratulate USAID for its role in providing food to 26 million Iraqis. As the chairman has mentioned, some of our trading partners have criticized food aid in the current WTO negotiations and suggested that we go to cash-only donations. This is a dangerous path. As the director of the

U.N.'s World Food Program has pointed out, food aid is easier to trace than cash. World Food Program and private volunteer organizations, such as CARE and Catholic Relief can consistently follow food deliveries every step of the way to their intended recipients. Food aid can also be targeted and nutritionally designed to reach women and children, who are the most often affected by chronic hunger. Cash assistance often does not reach women and children.

In addition, food aid can be delivered quickly, as the operation in Iraq proved, feeding 26 million people and delivering more than 2 million tons of food in just 7 months. At its height, the operation was moving 1,000 tons an hour, 24 hours a day, 7 days a week.

Mr. Chairman, thank you for your leadership on this issue. I look forward to hearing from our witnesses.

The CHAIRMAN. Thank you, Mr. Stenholm. It is now my pleasure to welcome our first panel.

Mr. Andrew Natsios, the Director of USAID, was scheduled to be with us, but he was called to Haiti just yesterday, and we are very pleased to have in his place Mr. William J. Garvelink, Senior Deputy Assistant Administrator for Democracy, Conflict and Humanitarian Assistance with the USAID, Mr. Joseph A. Christoff, Director, International Affairs and Trade with the U.S. General Accounting Office, and Mr. H. Lee Schatz, Special Counsel for Iraq Reconstruction, Office of the Administrator of the Foreign Agricultural Service, U.S. Department of Agriculture.

Gentlemen, we welcome all of you. We will make note that your full statement will be made a part of the record and ask that you limit your remarks to 5 minutes, and we will start with Mr. Garvelink. Thank you.

STATEMENT OF WILLIAM J. GARVELINK, SENIOR DEPUTY ASSISTANT ADMINISTRATOR FOR DEMOCRACY, CONFLICT AND HUMANITARIAN ASSISTANCE, U.S. AGENCY FOR INTERNATIONAL DEVELOPMENT

Mr. GARVELINK. Thank you, Mr. Chairman, and members of the committee. It is a pleasure to be here today. I was to express the apologies of Andrew Natsios, as you noted, he is in Haiti, and other than that, he was planning to be here today, and I have his comments for the record, but we will make just a couple of observations about USAID's role in ensuring that the Iraq population was fed over the past 15 months.

Prior to the war, as part of the interagency process, humanitarian offices within USAID began preparing for a humanitarian crisis in Iraq and began planning how to guaranty that the Iraqi population continued to receive food supplies through the war and into the post-conflict period. The humanitarian crisis did not happen, but Iraqis, most of whom relied on government rations before the war, had to be fed. With the full cooperation of the U.N. World Food Program, Iraqi officials, the State Department, the Coalition Provision Authority and the Civil Affairs Officers of Coalition Forces, USAID ensured that 26 million Iraqis continued to receive their regular food rations, and they continue to receive them today.

Averting a food crisis in Iraq has been one of the U.S. Government's major successes in Iraq. By March 2003, USAID had secured the agreement of the World Food Program to assume the overall management of food distribution under the Oil for Food Program, using Iraq's existing Public Distribution System. For years in Iraq, each Iraqi citizen has received a monthly food ration. These rations have been distributed through a network of almost 45,000 food and flour agents. Rather than create a new system, the World Food Program revived this national network and worked with the Iraqis to feed the Iraqi population, and this system is still being used today.

Managing this national food ration system has been an enormous task. It requires almost 500,000 metric tones of commodities to be distributed throughout the country monthly. A fleet of thousands of private and public sector trucks make daily trips in and out of Iraq, hauling food overland to more than 400 warehouses and silos around the country. The 45,000 food agents pick up these commodities at these locations for final distributions to the Iraqi family. As the war began in March, the U.S. assumed that Iraqi families had 2 to 3 months food supply. To ensure that there would be food available after their supplies ran out possibly in June, we undertook several actions. USAID provided $45 million to the World Food Program to fund its mobilization of staff and initial trucking, warehousing and fuel contracts. USAID provided to the World Food Program $200 million in cash and $181 million in Public Law 480 title II commodities and Emerson Trust food supplies. These actions totaled $426 million.

On June 1, 2003, food distributions began as planned through the Iraqi Public Distribution System, and they have continued each month ever since. For more than a year, the World Food Program, Coalition Provisional Authority and USAID have worked with the Iraqi Minister of Trade to train staff and prepare the Ministry for the time when it would assume full control of the Public Distribution System, and now on June 30, the Ministry will take over the system, and the last commodities procured by the World Food Program will arrive in Iraq by September 2004.

We fully support the hand over of the Public Distribution System to the interim government of Iraq, but we remain concerned that insecurity and management challenges might lead to disruptions in the food pipeline. Delays in food deliveries not only threaten the vulnerable, they could also create further instability among the population. We will remain vigilant in the coming months and be prepared to assist the new Iraqi government, should it become necessary.

Mr. Chairman, I cannot end my comments without noting that 2004 marks the 50th anniversary of the Public Law 480 Food for Peace Program. Over the past 50 years, the U.S. Government has contributed more than $50 billion to finance more than 376 million metric tons of food in over 150 countries. More than 3.3 billion people worldwide have benefited from this program. It is a credit to the dedication and commitment of this committee and its members that the United States has been able to share its bounty with so many of the world's needy. We are all justly very proud of this program.

Thank you, Mr. Chairman.

[The prepared statements of Mr. Garvelink and Mr. Natsios appears at the conclusion of the hearing.]

The CHAIRMAN. Thank you, Mr. Garvelink.

Mr. Christoff, welcome.

STATEMENT OF JOSEPH A. CHRISTOFF, DIRECTOR, INTERNATIONAL AFFAIRS AND TRADE, U.S. GENERAL ACCOUNTING OFFICE

Mr. CHRISTOFF. Thank you, Mr. Chairman. Thank you for inviting GAO to this important hearing.

For several months, GAO has been reviewing the operations of the U.N.'s Oil for Food Program, and today, I will discuss our findings and observations on that program, and describe current and future challenges to achieving food security in Iraq. First, let me discuss the Oil for Food Program.

Under U.N. sanctions, Iraq was allowed to sell oil to purchase food and other humanitarian goods. From 1997 to 2002, the U.N. controlled over $67 billion in Iraqi oil revenues and issued $38 billion in letters of credit to purchase commodities. The program helped the Iraqi people by almost doubling their food intake over the first 5 years of the program. However, the Iraqi regime acquired over $10 billion in illegal revenues during this period. This included $5.7 billion in oil smuggled out of Iraq and $4.4 billion in surcharges on oil sales and illicit commissions on imported commodities. Oil was smuggled through Syria by pipeline, across the borders of Jordan and Turkey by truck, and through the Persian Gulf by ship.

The Iraqi government also levied surcharges against oil purchasers and commissions against suppliers of commodities. The surcharges were up to 50 cents per barrel of oil, and the commissions were 5 to 15 percent of the commodity contracts. Now how and why did these problems occur? First, the Oil for Food Program gave the Iraqi government the authority to negotiate contracts directly with companies that purchased oil and supplied commodities. The MOU between the U.N. and the Government recognized the sovereignty of Iraq in negotiating oil and commodity contracts. Iraq's control over contract negotiations was an important factor in allowing the government to levy illegal surcharges and commissions.

Second, U.N. member states did not enforce the sanctions imposed in Iraq. Jordan maintained trade protocols with Iraq that allowed it to purchase heavily discounted Iraqi oil in exchange for up to $300 million in Jordanian goods. Syria received up to 200,000 barrels of Iraqi oil per day, in violations of the sanctions, and oil smuggling also occurred through Turkey and Iran.

Third, it is unclear how the United Nations assessed the reasonableness of the prices negotiated between the Iraqi government and commodity suppliers. In September 2003, the Defense Contract Audit Agency found that 48 percent of Oil for Food contracts were potentially overpriced by 21 percent. Food contracts were the most consistently overpriced. U.N. Sanctions Committee procedures stated that the Office of Iraq Program was to examine each commodity contract for price and value. However, OIP officials stated that no U.N. resolution tasked them with assessing the price reasonableness of the contracts. And the Sanctions Committee was responsible for approving commodity contracts. However, it primarily screened contracts for dual-use items rather than for price.

The United Nations and the Iraq Supreme Audit Board have begun investigations into the Oil for Food Program. These investigations offer an opportunity to determine the extent of the corruption, the adequacy of internal control and ways to improve the delivery of humanitarian assistance under economic sanctions.

Let me turn briefly to the challenges that Iraq faces in achieving food security. Sixty percent of Iraqis rely on monthly food rations as their primary source of nutrition. Ten percent are extremely poor and need additional food aid. However, the transfer of the Public Distribution System from the United Nations to the CPA and then to the Ministry of Trade has not gone smoothly. CPA's failed plans to privatize the system and delay negotiations with WFP resulted in localized food shortages in early 2004. Coordination between WFP and the Ministry of Trade has deteriorated. The Ministry has not regularly provided WFP with complete and timely information on food plans, stock reports or cargo arrivals. Furthermore, the worsening security situation has limited food deliveries from neighboring countries. Port congestion and few drivers willing to transport food have reduced the movement of food to warehouses. As a result, food stocks are tenuously low and the country has less than a 1-month's supply of several food items.

For the future, the Iraqi government must decide whether to continue, reform, or eliminate the public distributions system. The system is expensive and accounts for 25 percent of Iraq's budget. So, in conclusion, as investigations of the Oil for Food Program progress, the Iraqi government must take action to restore food stocks, target food to the most needy and procure food in an open and competitive manner.

Thank you, Mr. Chairman.

[The prepared statement of Mr. Christoff appears at the conclusion of the hearing.]

The CHAIRMAN. Thank you, Mr. Christoff.

Mr. Schatz, we are pleased to have your testimony.

STATEMENT OF H. LEE SCHATZ, SPECIAL COUNSEL FOR IRAQ RECONSTRUCTION, OFFICE OF THE ADMINISTRATOR OF THE FOREIGN AGRICULTURAL SERVICE, U.S. DEPARTMENT OF AGRICULTURE

Mr. SCHATZ. Mr. Chairman, members of the committee, I appreciate the opportunity to review the work of the Department of Agriculture in Iraq over the last year and assess the outlook for U.S. agricultural exports to that market.

What I would like to do this morning is provide a brief overview of Iraq's agricultural production, highlight the trade opportunities ahead for U.S. exporters, and also bring you up-to-date on the activities that the Department and our private sector cooperators are already doing to renew our relationships with Iraqis, and it truly is to renew those relationships.

I arrived in Iraq in April 2003, shortly after the fall of Baghdad. I spent 9 weeks at the Ministry of Agriculture as the first senior advisor, trying to bring that Ministry back up online. Since my return to the U.S., I have worked full-time for the Department on

Iraq issues across the board. We have had an additional 10 USDA employees working in Iraq on either short-term or longer-term assignments over the last year, and additionally, we have already identified the two foreign service officers who will be our first staff in the new Embassy in Iraq.

It might be useful to clarify the differentiation between the two Ministries we work with. That is the Ministry of Agriculture and the Ministry of Trade. The Ministry of Agriculture in Iraq focuses on supporting production agriculture in that country for food, fiber, livestock and poultry. The Ministry controls and administers land ownership, manages the water distribution to the farmers and proposes domestic procurement prices. One of the biggest activities of the Ministry is to import all production inputs, from seeds and fertilizers to fan belts for tractors for further sale onto farmers. The Ministry of Trade's focus is importing all the food distributed through the Public Distribution System, which is essentially the sum total of the market there.

Decades of State intervention and the economy have marginalized private market-drive initiatives in agriculture. For the past 20-plus years, Iraq's agricultural sector has effectively been cut off from innovation. The world's adoption and adaptation and use of high-yield varieties, modern herbicides and pesticides, the full range of improved production practices and new post-harvest technologies have largely bypassed Iraq. But despite that extended mismanagement, this is a country that has the resources, land, water and people, to still have a very successful agricultural sector.

However, if agriculture is to flourish, it will take time and require new leaders that will make tough decisions on changes ahead. Regardless of production gains, for the next several years, and I believe much longer, Iraq will rely on imports to meet a large portion of its food needs. To meet those food needs, the Ministry of Agriculture imports and distributes nearly half a million tons of food every month. We expect them to be the biggest purchaser of food over the next 12 months, at least. In that period, those purchases are going to include 2.6 million tons of wheat, a million tons of rice, 300,000 tons of pulses, 400,000 tons of cooking oil, 600,000 tons of sugar and a number of other commodities.

We have a potential for greater domestic production in Iraq and continued huge flow of imports. We have them both. Now if we look at just the growth potential in that market for one product and that one product will be wheat, we see that the per capita wheat consumption in Iraq today is 60 percent the level of Turkey, 70 percent the level of Iran, and only 80 percent the level of Syria. People right next door in the neighborhood, very similar systems. We need to remember that what these people have been receiving for 7 years is a food ration and not what they would choose to purchase for their food use, if they had a growing economy. In the long run, Iraq will remain a major commercial food market and a market that will demand higher quality from the importers who are supplying that market.

Since late last year, USDA has begun work, even with the situation in the country, with our private-sector cooperators in Iraq. That has meant most of the activity is outside of the country. The

U.S. Grains Council, the American Soybean Association have received USDA funds to help restart and grow Iraq's broiler and layer industry and modernize that industry out of its 1980 doldrums. We have also supported representatives of the U.S. wheat, rice and pulse industry to meet with Iraqi buyers in Jordan earlier this year. The result of just that first meeting with these people was to identify specifications that for the first time in a number years allowed U.S. products to be offered against Iraqi tenders.

We have also used the Foreign Market Development funds to hire a consultant in Iraq to begin looking at opportunities for other cooperators. Supporting and guiding market promotion activities will be if not the key focus, a key focus of our new team in the Embassy. We are currently working to reschedule a team of Ministry of Trade, decision-makers to visit the United States and will keep your committee informed on our progress on that.

Thank you.

[The prepared statement of Mr. Schatz appears at the conclusion of the hearing.]

The CHAIRMAN. Thank you, Mr. Schatz. Mr. Garvelink—as I mentioned earlier, Administrator Natsios could not be here. Please extend to him our congratulations on the upcoming 50th anniversary of the Food for Peace Public Law 480 Program. This program has aided 3.3 billion hungry people all across the world during its 50-year history.

As you are aware, the European Union and others have proposals in the World Trade Organization negotiations to end the U.S. Food Aid Program as we know it. What is your view on the future of food assistance around the world if the U.S. programs are not in place?

Mr. GARVELINK. I think the first implication of what would happen is people will die. 840 million people require food aid around the world every day, and to meet that need, I think about 8 to 10 million tons of food a year are provided by the international community. The U.S. Government provides 4 million tons of that total. Without that 4 million tons, it would be virtually impossible to even approach meeting the needs of the hungry people around the world. We provide about 56 percent of the food to all WFP appeals and we cover the majority of their operating costs through our contributions.

If our program didn't exist, WFP would have to dramatically scale back its programs, and again, hundreds of thousands if not millions of people would not be fed. I think in all honesty, our European Union colleagues are kind of long on talk but short on action. If you look at a couple current crises today, Ethiopia and western Sudan and Darfur, our European colleagues have not come through with the food assistance that is required, and if it had not been for our Office of Food for Peace and USAID and Ethiopia and presently today in Darfur, more people would be dying than already are in those countries.

Moreover, most of the countries who require our food aid are fragile states and are very much in danger of becoming failed states, and as we all know, failed states are a large breeding ground for terrorists and other elements that would do harm to the United States. I think our Food Aid Program with food aid is critical first of all to feeding millions and millions of people each year, but it is also critical to our national security, as we help fragile states emerge from that and not fall back into state failure. So I think our Public Law 480 program as it is, a food program is critical for the world's hungry and for the United States.

The CHAIRMAN. As for food aid, the United States' work in Iraq is perhaps the best example of how food aid can contribute——

Mr. GARVELINK. Correct.

The CHAIRMAN [continuing]. To a solution to a major, major problem.

Mr. Christoff, what countries on the U.N. Security Council were the most vocal about giving Saddam Hussein the benefit of the doubt after the U.S. and Great Britain began to notice that some of the Oil for Food contracts did not look legitimate?

Mr. CHRISTOFF. Mr. Chairman, I wish I had the minutes of the Sanctions Committee hearings, those are not available, to really understand who expressed what concerns. But I can tell you which countries benefited from the Oil for Food Program through the commodity contracts, and those key countries were Russia, France, Egypt, United Arab Emirates, Syria, Turkey and more recently China and Vietnam with the rice imports.

The CHAIRMAN. Thank you.

Mr. Schatz, despite its poor performance during the 1990's, Iraq's agricultural sector is known to have an abundance of agricultural resources compared with its Middle Eastern neighbors. This list includes fertile land and water from the Tigris and Euphrates rivers. A Saddam government guided by cronyism, tribalism and political patronage contributed to low agricultural productivity and a high degree of import dependency in the past. But several other factors also contributed to low agricultural productivity, such as disputes over land ownership, water rights and lack of research and technological advancements.

Many of these problems still exist. How difficult will it be to remedy these problems in Iraq's agricultural sector, and what can the USDA do to help advise the Iraqi Mnistry of Agriculture on these problems?

Mr. SCHATZ. I think the core problems are some things that take for granted; is land ownership and water rights, be it ground water or surface water. These are right at the heart of the problem because there is uncertainty in that system when you plant, and we saw it last year in the fall, if you are going to still own that land by spring. The inability to have clear title to land and water rights has really created a situation where there is no credit available. The answer to that was subsidized inputs to farmers to at least keep them moving.

I think the problems are essentially endemic of the system, and that in itself, quite frankly, to use the word system is part of the problem. The divide and conquer, command and control, if you will, mentality of the regime has made it impossible for anyone to really be responsible for enough of their own enterprise to be very optimistic about it. When you talk to people in the Ministries, they have Western educations. They know how to do things right. They are anxious to do it. They need some upgrading in their skills because they realize they have missed 20 years of opportunities. But it is going to be changing a couple of generations of a way of doing benefits of the agricultural community, and as we all know that is a tradition that changes a little slowly. So it is going to a very long, uphill battle, with the best of intentions.

The CHAIRMAN. Thank you. The gentleman from Texas, Mr. Stenholm.

Mr. STENHOLM. When you were listing the countries that benefited from the oil base, I didn't hear you mention Australia. Is that just an oversight?

Mr. CHRISTOFF. Australia was the chief procurer of exports for wheat, in particular. So yes, that was an oversight.

Mr. STENHOLM. I thought that was the case. It is interesting, when you say that there were overcharges of 22 percent, some countries can forgive their debt, and they are really just forgiven what they overcharged during that program. Weren't they?

Mr. CHRISTOFF. Could be.

Mr. STENHOLM. Pretty good deal. One thing you said that I think is really alarming looking forward—now not concentrating on what may have gone wrong in the past, but if Iraq has a 30-day supply of food, then I hope somebody is doing some planning as to what is going to happen to continue to feed the Iraqi people after June the 30, under what could be some very difficult situations, just as this morning we find that there their oil flow has been disrupted for at least 2 weeks now through bombings. That means that the potential income to pay for food in the new government is going to be less than it would have been otherwise.

What are the world's plans, what are our plans now to assist the new Iraqi government with the food that they are going to need until they can get their feet on the ground, so to speak? Anybody know? Mr. Schatz?

Mr. SCHATZ. I will take that one. Well, as someone who lives with this day in and day out and there are another couple here in the room that for the last year have watched this situation really closely, while 30 days doesn't sound like much, I am glad we are not 3 weeks ago when we were down probably to about a 2-week pipeline in the country. As the violence spiked about 2 weeks ago, all the trucks went off the road. CPA officials on the ground were instrumental in putting incentives to get those trucks back on the ground, again to move more wheat out of Umm Qasr. Since the spring, we have increased the capacity of that port to unload Panamexes. We had quite a demiurge problem there earlier this year. That has been overcome.

At the same time, we have within the last 10 days entered a period, which will run through September, when food supplies purchased earlier this year by the World Food Program begin to come in. World Food Program is responsible for delivery to the final warehouses, so they are using a lot of trucking from outside the country. First U.S. wheat is coming in at this point in time under the World Food Program contracts. It is coming in with trucks from the Jordanian and Syrian side. Through September, supplies should increase somewhat. There have also been some local purchases made of flour in the region to fill as a stop-gap measure. In addition, within the next 30 days, we would expect the purchasing by the Ministry of Trade to fill needs through the end of the year, to begin and be completed.

There is funding available with the Ministry of Trade for the remainder of the purchases needed to meet their food aid commitments under the Public Distribution System through the end of the year, and they will be forming their budgets for that system for the next 2 months.

Mr. GARVELINK. If I could just add to that, I think that USAID has also been in constant contact with the World Food Program. They have a contingency plan together that will deal with the most vulnerable portions of the Iraqi population. We have our food experts in-country that will be in, along with USDA, in the new Embassy that is there, and they are available to provide technical assistance of all different kinds, from food management to procurement to developing internationally-acceptable standards, negotiating procedures, documents, that sort of thing.

So I think everybody is keeping a very close eye on how these next few months will evolve when there is an overlap between what the Ministry of Trade does and the last 1.6 million that has been procured by WFP and will be coming in through September. So I think everybody is going to be very careful and ready to help, if need be.

Mr. STENHOLM. Yes. Final question. Mr. Schatz, you certainly have had experience on the ground, but when we talk about food aid commercial production, there is always that fine line between the aid program being disruptive to the normal market channels, and in my opening testimony, I indicated my strong desire that we look at ways in which we can regain some market share of wheat, rice in this country, using the same tools that the opposition use, that our competitors use.

Realistically, if you could just look at the next year, 365 days, what percent of the total consumption needs of cereal grains are going to be able to be financed through normal—although they wouldn't be normal. They would be abnormal. But commercial sectors, and what percent is going to have to be food aid in Iraq? Any guess on that?

Mr. SCHATZ. I guess I would offer an opinion.

Mr. STENHOLM. Yes. That is all I am asking is opinion.

Mr. SCHATZ. We know that for the next 6 months, they do have adequate financing. I think at this particular point in the new government's just getting started, the importation of food is going to be one of the top priorities. I think that is one of the dilemmas we have in wanting to dismantle this public distribution system, but at the same time realizing we already have some vulnerable populations and we don't have an economy that is broadly taking off. I think it is going to get priority for funding. Will it be able to acquire enough funding for the full year without beginning to look for other options, short-term credits, those type of things, I think is something they are going to be thinking about in 2005.

Mr. CHRISTOFF. If I could just add to that, it is going to be a difficult challenge for the Iraqis just because of the costs associated, as Lee mentioned, with the this Public Distribution System. The Iraqi budget is roughly $19 billion a year. The Food Distribution System is going to cost up to about $3½ billion to $4 billion to sustain this monthly ration system. So it would have to a top priority, and along with that is trying to understand what do you do with this Public Distribution System? How do you target the food to the most needy? If 60 percent of the population were dependent upon it, there still were 40 percent that did not use the entire food basket. So how do you get down to that 10 percent that are extremely poor and most vulnerable, and how do you manage a very costly distribution system that is going to affect your future budget?

Mr. GARVELINK. Just to add one more element of the perspective there. In some of the studies we have seen, the destitute and the term is critically poor population, that comes to roughly 20 percent of the population in Iraq, and that—by critically poor, the definition is less than $6 per year per capita of a family, and these families right now—when you are in the public distribution system, there is a nominal cost of 250 dinar to get your monthly food ration. That is 18 cents. And it is very difficult for some of these families even to come up with that, so there is a real need to target this system to make sure that the most vulnerable portions of the Iraqi population are not left out, as the transition gets underway at some point toward a market economy.

The CHAIRMAN. The gentleman from Nebraska.

Mr. OSBORNE. Thank you, Mr. Chairman, and thank you for being here today. The first question I have has I guess been alluded to one way or another, and I believe Mr. Christoff, you indicated roughly $10 billion was skimmed from the Oil for Food Program, something like that. It was a fairly sizeable amount, and I know that you realize where many of the trading partners were. I think you mentioned Russia, France, Egypt, Australia; several other countries. Is there any likelihood that any of that money will ever be recovered? Because I believe that under the Food for the World Program, Tony Hall, one of our former Members ran it, and he was indicating that maybe as high as 90 percent of the food distributed was through Public Law 480 and other programs from the U.S. So obviously, of that $10 billion, United States has a huge stake, and this is probably not the appropriate venue. But does anybody here have any conjecture as to whether any of that money will ever be adequately accounted for or recovered in any way from any of these trading partners?

Mr. CHRISTOFF. Part of the investigations I think that are occurring, the U.N. investigations is to begin looking at the actual contracts. There were about 30,000 contracts that were signed under the Oil for Food Program. I think one of the most important things to do is to look at which were the predominant companies that received contracts, which were the countries that those contracts came from, and try to get a sense of the potential overpricing to at least get a handle of the universe out there. Where should we target any future investigations? When you are also talking about where monies from the former regime might be hidden, the Treasury Department is heading up a 20-agency—interagency taskforce called the Iraqi Assets Working Group that is trying to determine exactly where former regime's assets have been hidden, and they have, through their efforts, recovered about $959 million of the former regime's assets out of an estimate of about $4½ billion. So there are efforts beyond the Oil for Food Program within the U.S. Government to try to return some of the former regime's assets.

Mr. OSBORNE. Well, thank you, and I think if you look at the palaces, more than 70, and we looked at the Water Palace which had to be hundreds of millions of dollars, I would assume that some of the assets are right there in plain view, and——

Mr. CHRISTOFF. The Mercedes are still over there, I am sure.

Mr. OSBORNE. Yes. Right. I have another question here. I think this would be for Mr. Schatz. We have heard a lot from some dry bean producers who are interesting in resuming trade with Iraq, and they are having trouble getting registered with the Iraq Ministry of Trade, and I guess if you don't get registered with the Iraqi Ministry of Trade, you don't do any business. And can you amplify or explain to us what the process is, what the holdup might be, or what the likelihood of any success might be in this regard?

Mr. SCHATZ. Well, I haven't dealt with that issue since communicating with both the industry in Baghdad since about 8:30 this morning. We have four companies from the dry bean industry who have at this point in time submitted all documentation that they were requested to submit to Baghdad. The farm and agricultural—or the Armed Service Agency has an individual actually working with the Iraqis who has been there 2½ months, trying to help them develop a system of registration of valid suppliers. We have been hand-carrying these documents to the Ministry, and like I say, at 8:30 again this morning, we have rattled some cages over at the Ministry, saying look, we want either response or do you need additional information? What we have recommend to the Ministry is that anybody who has submitted all documentation, if they have not received a response by the time a tender is let in, we anticipate them really soon, they would still receive those tender offers. Our goal all along is to make sure that our people have an opportunity to bid.

The difficulty we have found time and again over the last year is that unless it is a crisis, many of these Ministry people simply don't take action. They focus on what is hottest on their plate at that moment, and we don't see anyone stonewalling. We think it is just simply inaction as they focus on other items right now. But we are confident we are going to get that work through before the tenders are released.

Mr. OSBORNE. One last question. I know that MAP Foreign Market Development, Public Law 480 has been very important here. Do any of you see problems with WTO regulations here? I know that has been subject to some scrutiny, the $19 billion amber box limitation, and are any of you qualified to reflect on that, as to whether any of these programs have to be ruled against, or are they on sound footing? Nobody knows?

Mr. CHRISTOFF. I don't have a comment on that.

Mr. SCHATZ. No.

Mr. OSBORNE. OK. Then I am done.

The CHAIRMAN. Thank you, Mr. Osborne. The gentleman from Georgia, Mr. Scott, do you have questions?

Mr. SCOTT. Let me ask this question with respect to U.S. efforts to revitalize Iraq's agricultural sector. To what extent do you feel progress in rural areas is being hampered by the ongoing security concerns? I am very concerned about that. I serve as the co-chairman of our Democratic Group on National Security, and I would be interested to get your response on that. It would be Mr. Garvelink? Is that right?

Mr. GARVELINK. Yes. Well, I think security obviously is a problem and it has slowed down activities but has not brought them to a standstill. I think particularly in the areas around Basra where we are involved with the Marsh Arabs and trying to re-stimulate agriculture, their livestock, fish farming and that sort of thing, we have been able to do a considerable amount of work. In other parts of the country, of course, it becomes a problem.

Mr. SCOTT. Do you foresee any additional challenges in view of the upcoming June 30 hand over, this administration? Are things in place? Do you feel any apprehension because of that?

Mr. GARVELINK. Well, yes. The security situation is expected to get worse in the coming weeks and that will have an impact on all of our development activities, but hopefully we can get through that and get back on-track with the new government and sort things out and begin to move forward again on our agricultural programs.

Mr. SCOTT. When last were you over in Iraq, or were you over in Iraq recently?

Mr. GARVELINK. The last time I was there was probably late March or early April.

Mr. SCOTT. Do you, just from your opinion as a relatively recent person going over there, what is your assessment of our efforts in winning the hearts and minds of the Iraqi people, given the fact that you are involved in a very important part of that infrastructure agriculturally? Is your assessment that we are making progress in that area, with the recent situation evolving, do you think we are retrogressing in winning the hearts and minds of the Iraqi people?

Mr. GARVELINK. Well, obviously it will be a bit of a guess because I haven't been able to travel throughout the country. I was only in the central parts and them up in the north in the Kurdish areas. But my sense is a lot of the reporting we hear, which is natural, is on the military situation, and it is focused in Baghdad and Sadder City and other places. But I think if you travel more broadly in Iraq, you will find that a lot of people are very happy with what is going on. They are very appreciative of the types of assistance that the U.S. Government is providing, and it is a little bit different environment than we might see on a day-to-day basis on television. At least, that is what struck me when I was traveling around.

Mr. SCOTT. Yes. Well, thank you. Mr. Chairman, one other point?

The CHAIRMAN. Yes, please.

Mr. SCOTT. When I was over in the Middle East and visited in Israel, and I assume that some of the geography and the climate is the same, in terms of agriculture, they were making great gains in the handling of the distribution of water in that very dry, arid area. Is the irrigation systems similar to what they are using in Israel, and are they handling some of the same challenges there, or are you familiar with the——

Mr. GARVELINK. Well, I am not familiar with the system in Israel, so I can't really address that.

Mr. SCOTT. OK. All right. Thank you, sir.

Mr. SCHATZ. In Iraq, there is a massive program started actually under Oil for Food. We think of it as a food program, but it was much more of that. All of the inputs going into agriculture were also part of the Oil for Food-funded program. The Iraqis had undertaken in the last 4 or 5 years a massive program to switch from surface irrigation to more drip irrigation in their orchards, more center pivot irrigation, starting to tap their groundwater resources to get better efficiency, because one thing they have is a problem with salinity, same thing in Israel.

Mr. SCOTT. Yes.

Mr. SCHATZ. So they are beginning to try to use that technology. The dilemma is it is unclear exactly how good the planning process was to decide to start mining a groundwater resource in the edge of the Najaf Desert. So part of the difficulty with the development is we see snapshots of things going on. I traveled in 9 weeks more extensively than my replacement, who was there for 8 months because of the security situation. So it is very hard to get a feel broadly for what is happening in agriculture. At the local level, we are having some great progress. But to knit together a national system that is sustainable from a national budget point of view is a very big challenge right now with the security situation.

Mr. SCOTT. Yes. I can imagine. That is my opinion too. Thank you very much. Mr. Chairman.

The CHAIRMAN. Thank you, Mr. Scott. The gentleman from Kansas, Mr. Moran.

Mr. MORAN. Mr. Chairman, thank you. Could you paint for me the picture of what the size of the food market will be, agriculture commodities currently and in the future will be? What amount of money, the size of this market?

Mr. SCHATZ. Import market, or——

Mr. MORAN. And the follow-up question is and that what percentage of it is being provided domestically, and then what percentage will be an import market?

Mr. SCHATZ. Well, let us just take wheat for a minute, and we will jump to rice.

Mr. MORAN. Well, wheat is a good one to take.

Mr. SCHATZ. I sent a message to the Wheat Commission this morning actually in Kansas, some information. Right now this year, we will probably procure in the neighborhood of half a million tons of millable quality wheat out of the Iraqi crop. Iraqi crop this year, wheat may be 2 million tons, a problem is basic agricultural statistics. This is a State planning country, so we don't really know. We are not able to get out and survey. Last year, there was about I think 800,000 tons we purchased, millable quality.

This year, we worked to get a lot of pretreatment chemicals out to the seed people to try to get the smut down, so we think we are going to have a better crop. That 500,000 tons is going to be mixed with about 2½ million tons of imported wheat to provide a total wheat availability that is going to be 60 percent of Turkey's. If we got that production up double in the country, there is still the demand there if we have got an economy going. The Australian advisors and the American advisors have agreed all along we are not afraid of a more productive agriculture sector, as long as there is a better economy in the country, because this is a society that is still going to eat more food.

If you were to drive through the streets of Baghdad right now, first it would be more dangerous than it was a year ago. Secondly what you would be impressed by is all the fresh fruits, all the vegetables you are going to see in the marketplace. There is an abundance of fresh produce, fruits in the marketplace throughout a very extended season, starting in Basra in the south, going all the way to the Kurdish areas in the north that markets actually through a private-sector channel. So when we say OK, it is $3 billion of food imports, we can't put a figure on how much the cucumber crop is, how much the tomato crop, the onion crop—but had I had to live off the local economy a month after the war, I could have done it very easily, if I had the money to pay for it.

Mr. MORAN. Are the fruits and vegetables domestically grown?

Mr. SCHATZ. They are domestically grown. A lot of work has to be done on some varietal improvement, some post-harvest technology. They are not the kind that you would want to pick up at your local grocery store and put in your bag, but they have got pretty good flavor.

Mr. MORAN. Does the distribution of those fruits and vegetables or other agriculture commodities transcend the regions, the tribes, the religions of Iraq?

Mr. SCHATZ. If I understand the question, do they move freely and long distances? They move to the major cities. We saw trucks loading as far north as Mosul, and those trucks—we saw the same ones later that evening all the way in Baghdad coming into the wholesale markets there.

Mr. MORAN. What other countries are involved in food aid and also in developing a market for their agriculture commodities? Who are our competitors in Iraq, both in providing food aid and ultimately in selling agriculture commodities to Iraq as the economy improves?

Mr. SCHATZ. I think the three toughest customers we are going to have are Thailand and Vietnam on rice. Their rice quality has improved a lot since our product has been in there. Iraqis remember our product. They prefer it. We have got quite a premium right now because of our large exports earlier this year. Australia is going to be a tough competitor. They have got a lot of flexibility in that region. They have already got a lot of major customers. It is in their back door. They have got a freight advantage. I think those are the big three that we are going to see.

Mr. MORAN. My time is about to expire. You mentioned visas and trade officials visiting the United States. What success have we had of that occurring? I know in our case in Kansas, we thought we had Iraqi trade officials visiting, and visas were not issued for the entire party and the trip didn't take place. Is that commonplace, or is that an anomaly?

Mr. SCHATZ. It is difficult to get any people into the U.S. now compared to what we were familiar with a couple of years ago. What happened in this case is at least one of the visas in a package of visas got separated. I can report though at this point in time, everyone on that entire team has received their visa and we are currently looking at rescheduling that trip hopefully by mid-July.

Mr. MORAN. Thank you, Mr. Chairman.

The CHAIRMAN. Thank you. The gentleman from North Dakota, Mr. Pomeroy.

Mr. POMEROY. Thank you, Mr. Chairman. While I have my subcommittee chairman here, I would just like to pose a question to you on another matter. The RMA is redoing the Standard Reinsurance Agreement. This was a provision that I added into the ARPA legislation two terms back. There have been some controversy about the terms of the renegotiation, and naturally, if they get it wrong, it is going to impact significantly the effectiveness of this legislation. I hope we will be having an oversight hearing on these Standard Reassurance Agreement discussions underway and would ask either the chairman or subcommittee chairman if any are planned.

The CHAIRMAN. If the gentleman will yield?

Mr. POMEROY. Certainly.

The CHAIRMAN. I will tell the gentleman that we share his concern about that being done correctly. I have joined with the ranking member of the committee and others in writing to them as they have negotiated this, and we have made some progress in terms of getting them to look at this from a realistic standpoint. We are very concerned that there be both efficiency in the administration of the program and that there be fairness to the array of different companies that are engaged in the process.

So we will continue to examine that, and while I cannot pinpoint exactly the course that our oversight will take, we are definitely interested in examining this closely, and I would certainly welcome the possibility of holding hearings on the matter. I don't know if the gentleman from Kansas has scheduled any on the subject, but we certainly are watching it very closely.

Mr. MORAN. If the gentleman would yield?

Mr. POMEROY. I yield.

Mr. MORAN. Mr. Pomeroy, our subcommittee does have a hearing scheduled in July concerning crop insurance. Not specifically the RMA, but a follow-up to the series of hearings we have held in the past on crop insurance improvements, and knowing your capabilities, I know that you will be there with the opportunity presenting itself to ask questions of RMA on a topic that is slightly outside the topic of the hearing. So you will have an opportunity to discuss this topic with RMA officials in July at a hearing on crop insurance. We are also, as the chairman says, monitoring the negotiations, seeing the reaction, listening, paying attention and certainly willing to work with you and Chairman Goodlatte to make sure that our Subcommittee does what is appropriate to see that these parties come together to provide crop insurance for American farmers.

Mr. POMEROY. Reclaiming my time, I thank both chairmen. I have been not hesitant about giving RMA my own opinions as this process has gone on, and I am pleased to hear the Chairman say maybe a hearing would be held sometime in the future. Hold open the possibility anyway because sometimes, it is useful in one forum to get it all out, all views and just help us sort it out. It is an oversight function I think appropriately undertaken. Again, if they get it wrong, the Bill that we pass isn't going to work as well as intend it. And so it is a pretty direct stake here for us.

Mr. MORAN. If the gentleman would yield?

Mr. POMEROY. I yield.

Mr. MORAN. It has been my opinion that we want RMA and the insurance companies to negotiate, and that Congress does have a significant role to play in Oversight, but let us let the parties work their selves through the process before we become parties to those negotiations. But I welcome the opportunity to make certain that the outcome is something that is advantageous to the crop insurance—the beneficiaries, the folks that you and I represent in North Dakota and Kansas and across the country who desperately need that program.

Mr. POMEROY. Reclaiming my time, I think that that is an absolutely legitimate rationale to watch. We are now on draft number three though, and maybe at some point, we are getting closer to where we ought to take a look.

Now to the matter at hand, and I see my time is about to expire. First a couple of comments. To USAID, what you have achieved I think is remarkable. The logistics that under-gird your work with World Food Program are fantastic, and in the end, it is the result of work under dangerous conditions by really dedicated and talented people. Two that I know of are Regina Davis, as well as the National Guard Captain Devina French, who during her term there was very involved with this in Baghdad, and there are an awful lot of others. But, just as an example of the heroic performances that have kept people from starving, and you really have done I believe the United States proud in the operation of your agency.

Mr. GARVELINK. Thank you.

Mr. POMEROY. Now to Mr. Christoff, while we do with relief note the success of our operations at keeping things from having people starve so far, I am concerned about your observations about food stocks being dangerously low and systems really pretty well shot, as we look at the hand over coming up at the end of the month. Is this an area where we should entertain very significant ongoing concern relative to this working and did we basically blow a lot of the last year on a failed effort to try and privatize and get a different system in place? Yield.

Mr. CHRISTOFF. Well, I think seeing as some of the efforts back in November to privatize did delay the procurement of certain food stocks and did cause some local shortages, but certainly over the course of this entire year, Iraqis were still getting their monthly food rations. Right now, the security situation is pretty bad in Iraq, and therefore, you have food that is at the port that can't get to the warehouses, on the borders that can't get to the warehouses, drivers that are somewhat unwilling to drive in an insecure environment. So as I am hearing USDA and USAID say that vigilance is importance to keep track of where the food stocks are in this unsecure environment, and yes, it is important to keep track of this because food security—with 100 percent of the population having receiving these rations, it is viewed as a right. It is viewed as an entitlement, and it is viewed as something that is extremely important to the typical Iraqi family of seven who need the food on a monthly basis.

Mr. POMEROY. While the chairman is distracted, I will sneak in another point.

The CHAIRMAN. Without objection, the gentleman can have another minute to ask questions, since he did use some time on it with another worthy topic.

Mr. POMEROY. I thank the chairman. One of the observers of this whole matter opined we are fighting forces they don't have an alternative system they are trying to put in place. They just want wreck anything that is working, and so I do think it is a terribly daunting challenge, and are you seeing food stocks diminish? Is this the situation that is growing more alarming? Are kind of at a dangerous and unacceptable status quo, but it is holding in an equilibrium?

Mr. CHRISTOFF. Well, I think everyone is concerned about the time up to the transition. We have consistently heard reports about the insurgency, the foreign fighters, the increase in violence, the oil pipeline that was disrupted and it is going to be out for 10 days now, which is going to disrupt an important source of revenue for the Iraqi government. So it is important to try to have, as what I am hearing USDA say, contingency plans to try to deal with the future of the food security situation in Iraq.

Mr. GARVELINK. Thank you. Just to add a point, about 2 or 3 weeks ago was the most severe time when we were down to about 2 weeks supply because of the security situation, the difficulty of trucks moving, or their inability to move at all. I think the situation has improved in the few weeks, and now we are up to a month and we would like to get the buffer stocks a lot higher than that, so it is a very tenuous situation, but it seems to be improving a little bit right now.

The CHAIRMAN. I thank the gentleman. The gentleman from California, Mr. Ose.

Mr. OSE. Thank you, Mr. Chairman.

Mr. Garvelink, am I correct in understanding from Ambassador Hall that there are approximately 50 warehouses spread around the country in which these food stocks are secured?

Mr. GARVELINK. Do you mean for the Public Distribution System?

Mr. OSE. Yes.

Mr. GARVELINK. I think there is about 400 warehouses and silos scattered around the country in all 18 governance where the food comes in from the ports, its truck comes from Umm Qasr. It comes from Jordan and Akaba and Syria and Turkey and it goes to one of those, and then the local folks come and pick it up there. But I think there is about 400 of them.

Mr. OSE. Are the points of distribution secured?

Mr. GARVELINK. I can't really answer that question. I would suspect they are secure in some of the outlying areas in the north and south, and probably not as secure in and around Baghdad.

Mr. OSE. Well, I do want to compliment how AID and World Food Program have taken the average Iraqi caloric daily intake from 1,270-odd to 2,200. I think that has made an enormous difference in the survival rates of women and children especially in Iraq, and I want to compliment you for that.

Mr. GARVELINK. Thank you.

Mr. OSE. It is not a question. You don't need to make a statement. I am going to tell you don't make a statement.

Mr. Christoff, I want to make sure I understand some of the comments in your testimony about the Oil for Food Program. I notice in your scope and methodology, the second bullet point you used a discount rate of 67 percent for smuggling through Jordan and only 33 percent through Turkey, the Persian Gulf and Syria. Now I have some questions I am going to submit to you in writing. I presume the record is going to remain open for a couple days.

I want to go to Security Council resolution 986, which set up this entire process. If I understand correctly, the Security Council passed that Resolution, set up the Oil for Food Program and set up a Sanctions Committee to monitor the program. Am I correct in my understanding that the Sanctions Committee is the same as the Security Council?

Mr. CHRISTOFF. Yes.

Mr. OSE. So the same members who are on the Security Council are the same people on the Sanctions Committee?

Mr. CHRISTOFF. That is correct.

Mr. OSE. And the five permanent members of the Security Council include Russia and France?

Mr. CHRISTOFF. Correct.

Mr. OSE. All right. Now we are also a member of the Security Council, permanent member, so we are a member of the Sanctions Committee. Who at the Department of State represented the U.S. Government in the oversight of the 30,000 contracts for various items in the Oil for Food Program?

Mr. CHRISTOFF. The contracts were reviewed as part of a 60-person interagency review process that includes the Departments of Commerce, Energy, State and Defense as well, and the Intelligence communities to try to review all of the contracts primarily focusing on whether or not any of the contracts include items that could be used for military purposes, so called dual-use items.

Mr. OSE. OK. Someone somewhere had copies of these contracts?

Mr. CHRISTOFF. Yes. All members of the Sanctions Committee were given copies of the contracts for review.

Mr. OSE. How can we get copies of those contracts? How can Members of Congress access copies of those contracts? To whom do we need to send the request?

Mr. CHRISTOFF. The [House Committee on International Relations], for example, has sent a request to the U.S./U.N. Mission to try to obtain copies of those contracts. Yes. State has the contracts.

Mr. OSE. Well, we have been able to track through the Defense Contract Auditing Agency estimated 7,500 contracts, or which they have audited about 10 percent, so around 700. And of those 700 contracts, they have found significant unexplainable charges, much to in line with what you have what you have laid out here. I am trying to find out to whom would we send the specific request, either at Commerce or State or whomever, for this same information? Apparently International Relations has done the same thing. I understand why the U.N. would want the investigation done by

Mr. Volker to be done in England as opposed to being jurisdictional here in the United States. But to whom do we send the letter seeking copies of those documents?

Mr. CHRISTOFF. Well, we work through the U.S./U.N. Mission in New York, and that would be the vehicle by which to request contracts, copies of the contracts. Now the information though there is information at the U.S./U.N., at the U.N. There is also information in Baghdad that is being reviewed, in terms of the completeness of the contracts.

Mr. OSE. Well, there is an issue in Baghdad as to whether or not we can actually access those contracts there and who is going to provide them and who is going to pay for the research and so it is a different issue. But you are saying that the contracts that were reviewed numbering 30,000 the Sanctions Committee that the U.S. Government somewhere has copies of those contracts?

Mr. CHRISTOFF. Yes. The Department of State should have copies of them.

Mr. OSE. Thank you, Mr. Chairman.

The CHAIRMAN. The gentleman from Georgia, Mr. Burns.

Mr. BURNS. Thank you, Mr. Chairman. I appreciate the input from the committee.

I want to just get you to give me an update. I appreciate the various programs and the focus we have to be—is on making sure that these programs are administered effectively and that the U.S. taxpayers' value is achieved. But ultimately, the Iraqi people must feed themselves. When I visited Iraq in October and I saw their potential, can the panel comment on their ability to be self-sufficient in food?

Mr. CHRISTOFF. Well, they have never been self-sufficient in foods. Maybe the early 1980's or at best or even before that. So there is always the dependency on exports from all the agricultural producing countries in order to help Iraq with its shortfalls.

Mr. GARVELINK. Well, it would just seem from the little I have really focused on this that in prior years, agricultural production and its inputs have been so heavily subsidized on one side by the Iraqi government and peculiar policies have been put in place in the other that is has been very difficult for agriculture to develop over the past 15 to 20 years.

Mr. BURNS. In some areas of the north though, weren't there substantial irrigation projects that provided them the ability to provide certain supplies of grain?

Mr. GARVELINK. Yes. I think in the Kurdish areas in the north which have been protected by the no-fly zone for a number of years, I think they have flourished compared to the rest of the country.

Mr. SCHATZ. A large amount of that northern grain-growing area though is dry land.

Mr. BURNS. That is true.

Mr. SCHATZ. They have had now 3 years of good rain, but they had 2 years before that of just disastrous drought.

Mr. BURNS. Do you see their ability to contribute to their food supply as a potential impact upon our policies?

Mr. SCHATZ. I think their challenge is going to be undoing 2 generations of being paid for a product that is not a quality product. I mean, we had a lot of resistance a year ago. They just don't understand why we wouldn't take wheat that had 20 percent dockage in it, which is quite an amazing pile of wheat. It is going to take a change in really all the cultural practices. The thing I think we have to remember is when we look backwards, and you will hear people say that they were a surplus agricultural producer, they were exporting dates at a very high price at that point. That quality is down. Dollar value, yes. They were on that plus side. At the same time, they didn't have 27½ million people. That population, we heard the average family size is seven, so we know that we have got a steeply increasing population.

I think they can contribute a lot more, but I think what we are going to see is if that economy grows, they are going to demand better quality, and that better quality is probably going to come from outside the country. A year ago, I could buy Dole bananas on the marketplace in Baghdad, and they were the same quality Dole bananas I could buy here in the United States. This is a market that when they have some money is going to be a market for just about anything I think we can sell there that gives them a pointed difference. We are seeing product moving in already in greater quantities from Jordan, from the Emirates, as entrepreneurs are putting together essentially a grocery store container, taking it on a truck and selling that in the country.

So they can do better. I don't fear them doing better, as long as that economy grows, and I think it will because if you have a successful agriculture, you are going to have to have a successful country.

Mr. BURNS. Thank you. Thank you, Mr. Chairman.

Mr. OSE. Would the gentleman yield?

Mr. BURNS. I would be happy to yield.

Mr. OSE. Mr. Christoff, I didn't ask my question as thoroughly as I should have. Does GAO have copies of these contracts?

Mr. CHRISTOFF. Yes.

Mr. OSE. Have you seen the copies or the contracts themselves?

Mr. CHRISTOFF. We did look at some of the summary information on the contracts that had been renegotiated. These were the contracts that also were part of the Oil for Food Program when it ended and it was transferred. So we have some of those contracts but not the comprehensive contracts.

Mr. OSE. Do you have a list of the 60-odd people who were part of this——

Mr. CHRISTOFF. I would have to go back and check on that as to who exactly was part of the 60-person interagency review process. I could submit that.

Mr. OSE. I would appreciate that. Thank you, Mr. Chairman.

Mr. CHRISTOFF. My staff is reminding me of a few things in that. State Department has the lists of the contracts. The entire contracts are either with—definitely with the United Nations and also within Baghdad. But in terms of your desire to request the contracts, you would still be going through the State Department.

Mr. BURNS. Reclaiming my time, I would like to add my support for Mr. Ose's request to get the information that we need to make sure that the oversight and the execution of these contracts are reasonable for the American.

Thank you, Mr. Chairman. I yield back.

The CHAIRMAN. Thank you. I thank the gentleman.

Mr. Christoff, how satisfied are you with the organization and progress of the investigation that the U.N. is conducting in this matter?

Mr. CHRISTOFF. I am not really certain what progress the U.N. has made up to this point. I met with Mr. Volker a few months ago. He certainly is determined to try to get at the root cause of some of the problems with the Oil for Food Program. His first priority is to try to look at whether or not there were any U.N. personnel that might have been involved in inappropriate or illegal activities.

The CHAIRMAN. Let us assume that Mr. Volker comes forward with a report that identifies the culprits, and I think that is a fair word to use here, both in terms of individuals within the United Nations and individuals engaged in brokering these arrangements and individuals in Iraq and in other countries. How effective do you think the United Nations will be in taking action against its own employees and the others that are involved in what is obviously a major criminal activity?

Mr. CHRISTOFF. Well, I would hope they would take strong actions against any individuals that might have been engaged in inappropriate activities. We have been looking at this for 2 years. We started our work a couple of years ago, and there were no evidence that we found of any U.N. officials that might have done something inappropriately.

The CHAIRMAN. And finally, how confident are you that this will not occur in the future with this ongoing operation to provide food as needed in Iraq?

Mr. CHRISTOFF. And with the new Ministry taking over, I am hopeful that by having Inspectors General in place in the 26 Iraqi Ministries, having a Supreme Board of Audit, which is the GAO equivalent over in Iraq, and also WFP and others trying to teach Iraqis how to procure in an open and competitive environment, having codes of ethics and conduct that CPA has developed, I am hopeful that all of this will take hold and we will go beyond the corruption legacy that has existed within all the Iraqi Ministries.

The CHAIRMAN. I thank you. I expect that is a recess, but we will check that. Does any other member of the committee have a follow-up question or two they would like to ask? The gentleman from Texas.

Mr. STENHOLM. Mr. Christoff, if I understood what you just stated to the chairman that you looked at these contracts 2 or 3 years ago and found no evidence of misappropriation of funds by U.N. staff officials?

Mr. CHRISTOFF. We did not look at specific contracts, but what I am saying over the course of all of our reviews, we did not uncover any evidence that U.N. personnel did anything inappropriate or illegal.

Mr. STENHOLM. It strikes me in the GAO report where you now have found that middle-man review also found the use of middle-man companies potentially increase contract prices by 20 percent or more. Who are we talking about there?

Mr. CHRISTOFF. First of all, that is the Defense Contract Audit Agency's report and that is—this is—had been used throughout the Oil for Food Program in which you have one country that would negotiate the contract with Iraq and then purchase the commodities for another country. Russia is a good example. Russia often negotiated wheat contracts with Iraq and then bought the wheat from Argentina or Canada to then sell it. This consistent use of middle men, the Defense Contract Audit Agency found that it oftentimes would increase the overall contract price by 20 percent, and they recommended in the future that the use of middle men be avoided.

Mr. STENHOLM. Pretty good recommendation.

Mr. CHRISTOFF. Yes.

Mr. OSE. Would the gentleman from Texas yield?

The CHAIRMAN. I will yield to the gentleman from California.

Mr. OSE. The provision in the original resolution, correct me I am wrong, Mr. Christoff, was to preclude the use of middle men, and that is one of the fundamental things that Volker needs to look at, is that the original resolutions said you may not use middle men because they anticipated this problem. And in the process of implementing the program, that kind of got shoved aside and the middle men came into the debate.

The CHAIRMAN. I thank the gentleman. I would like to thank all of the members of this panel for your contribution on the ongoing situation in Iraq. We will continue to follow this very closely. At this time, we will move to our next panel. I would like to welcome the members of our second panel. Mr. John King, vice chairman, USA Rice Council of Helena, AK, on behalf of USA Rice Federation and U.S. Rice Producers Association, and Mr. Lochiel Edwards, president, Montana Association of Wheat Growers, Big Sandy, MT, on behalf of the Wheat Export Trade Education Committee, National Association of Wheat Growers and U.S. Wheat Associates. Gentlemen, we welcome both of you and advise you that your full statement will be made a part of the record and ask that you limit your oral remarks to 5 minutes.

Mr. King, we will start with you. Welcome.

STATEMENT OF JOHN KING, VICE CHAIRMAN, USA RICE COUNCIL OF HELENA, AR, ON BEHALF OF USA RICE FEDERATION AND U.S. RICE PRODUCERS ASSOCIATION

Mr. KING. Thank you, Mr. Chairman.

Mr. Chairman and members of the committee, my name is John King. I am a rice and soybean farmer from Helena, AR, and I am testifying today on behalf of both the USA Rice Federation and the USA Rice Producers Association. I am pleased to appear before the committee today on behalf of the rice industry. On a personal note, I will share with you how excited and honored I am to testify before Congress for the first time.

The U.S. rice industry wants to play a major role once again in supplying rice to Iraq. With the current challenges facing the U.S. rice industry, particularly in the milling sector, renewed Iraqi market access could have a tremendous impact in value-added sales and enhanced economies in rural America. In the late 1980's, Iraq was the number one market for U.S. rice. The United States provided nearly 90 percent of Iraq's rice imports, which totaled over 500,000 metric tons in peak years. In 1991, the U.S. Government exposed export sanctions, and this market was lost. Rice is no stranger to this situation, as Cuba and Iran were also once top export markets ultimately lost to the Government-imposed sanctions.

The impact of losing our number one export market has a significant negative impact on the U.S. rice industry and a positive impact on the rice industries of major competing rice export countries, primarily Vietnam and Thailand. As a result of the embargo on exports to Iraq and based on a U.S. share of Iraq's rice market in the 1980's, we estimate that the United States lost about $1.9 billion in rice export sales to Iraq from 1991 to 2003. The liberation of Iraq in 2003 by coalition forces has brought freedom to the Iraqi people. The resumption of trade has also provided hope for the U.S. rice industry. The Iraqi market has grown from about 550,000 metric tons in 1991 to 1.1 million metric tons in 2004.

USDA estimates that Iraq will need to import over 1 million tons of rice per year for several years. Reestablishing a viable commercial trading relationship between the U.S. and Iraq must be the goal of rice producers, processors, exporters and the Federal Government. To help facilitate this, leaders from the U.S. rice industry traveled to Amman, Jordan in February 2004 to meet the members of the Iraqi Grain Board in an effort to restore a trading relationship between the two countries. A follow-up meeting between the Iraqi officials and the U.S. rice industry was recently postponed, but are hopeful that this will be rescheduled very soon.

Market development will be vitally important in the months and years ahead, as U.S. agriculture attempts to reestablish its markets in Iraq. It is imperative that Congress provides full funding for MAP and FMD at levels authorized in the 2002 farm bill if sufficient funds are to be available to meet current needs and reestablish our markets in Iraq. It is critical that GSM export credits be made available for exports to Iraq as soon as possible. Some in the Administration have suggested that no GSM–102 financing can be provided to Iraq, since the former regime is in arrears. However, it is our understanding that current law has—understanding that under current law, having an outstanding debt or a rescheduling arrangement does not preclude USDA from offering a country a new GSM–102 credit.

There is no statutory impediment to prevent the Department from offering a GSM credit program to Iraq. Given this situation, we strongly recommend that the USDA announce its intention to establish a program for a GSM–102 credit for Iraq as soon as possible. This credit package is a critical, critical component to helping the U.S. rice industry rebuild its presence in this major market. We are sympathetic to the needs to address Iraqi's outstanding debt to the United States and we are committed to providing whatever assistance that we can to accomplish this goal. But we are of the strong opinion that Iraq's debt issue can and should be addressed concurrent with the provision of new GSM credits.

In summary, Mr. Chairman, Iraq was a top export market for the U.S. rice industry in the past, and we hope it will be again. We urge Congress to support Market Development Programs to help us rebuild relationships in Iraq, and we seek GSM credit assistance to facilitate future sales.

Thank you again for the opportunity to testify here today, and I would be pleased to address any questions that you may have.

[The prepared statement of Mr. King appears at the conclusion of the hearing.]

The CHAIRMAN. Thank you, Mr. King.

Mr. Edwards, we would be pleased to have your testimony.

STATEMENT OF LOCHIEL EDWARDS, PRESIDENT, MONTANA ASSOCIATION OF WHEAT GROWERS, BIG SANDY, MT, ON BEHALF OF THE WHEAT EXPORT TRADE EDUCATION COMMITTEE, NATIONAL ASSOCIATION OF WHEAT GROWERS, AND U.S. WHEAT ASSOCIATES

Mr. EDWARDS. Yes, good morning, Chairman Goodlatte, members of the committee.

I am Lochiel Edwards. I farm on the high prairies of Montana. My family and I work to raise high-quality milling wheat for the export market, as well as domestic mills. Today I am representing the Wheat Export Trade Education Committee, which is our trade policy arm in the wheat industry, the National Association of Wheat Growers, which handles domestic policy and farm programs, and for U.S. Wheat Associates, which is our foreign market development organization. I appreciate this opportunity to speak to you on where we now stand and what needs to be done to rebuild the Iraq wheat market.

I would first point out that we have in this country only 4 percent of the world's consumers, and we have a large wheat industry. We export 50 percent of our production yearly. We need markets to have a viable, ongoing wheat industry to provide a grassroots engine for our economic machine. As early as 1963, Iraq was an important customer of the United States wheat industry. There was a year in the 1970's when we had 100 percent market share. In most years leading up to the 1991 Gulf War, we maintained about one-third of the total annual Iraqi wheat import market at around 3 million tons a year. During the first Gulf War and the period following, Iraq refused to make any purchase from the U.S., not only because of the war, but also because of the unilateral sanctions we maintained against imports of their products.

Regaining market share is difficult at any time. The challenges in a war-torn country will no doubt add to the difficulties of regaining market share that has been displaced by our competitors. During the 1980's through 8 years of war between Iraq and Iran, Iraq was a high-priority market for our industry and U.S. Wheat Associates put a great deal of work into servicing this market. Dealing with the U.S. free market system was more difficult for Iraq than the government-to-government system they were able to use with the Canadian Wheat Board and the Australian Wheat Board. While not able to sign long-term purchase agreements with Iraq like our competitors could offer, the U.S. wheat industry signed a protocol with Iraq's State Organization for Grains, known as the SOG, encompassing technical cooperation. The agreement provided for an in-depth comprehensive program of technical assistance, trade servicing and information, in return for SOG's agreement to give the U.S. wheat market consideration as a major source of supply in meeting Iraq's annual imports.

By 1989, 1990, we allocated Market Development Funds for Iraq three times higher than the funding activities in Egypt, which is the region's largest wheat importer. This was done because the potential market in Iraq was growing quickly. Today with an opening of the market following the end of the Hussein regime, the U.S. wheat industry is working to renew friendships and pick up where we left off in offering all the assistance they need. We strongly believe that Iraqis and Americans alike benefit from the development of an open and competitive marketplace, where economies and specific end-use needs rather than political considerations determine what type of wheat is bought and from where. Open markets work when given the chance. They work best when commercial buyers are free to choose their sources of supply and when suppliers compete fairly and transparently.

The future for this market is positive. Iraqis are enthusiastic about once again working with us. The U.S. Wheat Associates was encouraged by a meeting with Iraqi's Grain Board, Foodstuffs Trading Board and the Ministry of Trade that was held in Jordan this year. They clearly wanted to learn as much as they could about getting back into the U.S. marketing system. We believe Iraqis want a competitive purchasing system and they want the U.S. to be a supplier and competitor. Iraq buyers have made a lot of progress in understanding wheat quality specifications, and we believe that will only get better as the Iraqis become reacquainted with the U.S. marketing system and the role of the Federal Grain Inspection Service in providing for official certification of quality at loading.

There have been roadblocks in recent attempts to bring Iraqi teams to the U.S. for training. One of these has been the difficulty in getting visas in a timely fashion, even when the individuals were invited and sponsored by the U.S. Government. Another difficulty is the lack of ability to extend USDA's credit program. Iraq's debt will need to be restructured so the country can begin to recover without the overhang of an estimated $150 billion. As I stated in the beginning of this testimony, it will take a lot of effort to reestablish U.S. wheat in this previously closed market. Our industry will do all that we can, but we need the assistance of the U.S. Government in the following areas.

We need to work with other countries in debt forgiveness and debt restructuring. We need to work on the allocation of GSM credits. Providing expeditious approval for visas for approved Iraqis to visit the United States is needed. Funding for market development activities and improvement in Iraq's port and grain-handling facilities is also a priority. Two more are travel authority to visit Iraq for market development activities, and the U.N. and other bodies are investigating kickbacks of the previous regime. We would ask that the U.S. Government firmly support that effort.

Mr. Chairman, I am very pleased to be able to report that Iraq has purchased U.S. wheat this year. As the Iraqis regain their familiarity with the U.S. marketing system and rediscover the unique qualities of U.S. wheat, we are confident we can recapture this market. So thank you, Chairman Goodlatte, members of the committee.

[The prepared statement of Mr. Edwards appears at the conclusion of the hearing.]

The CHAIRMAN. Thank you, Mr. Edwards.

Let me ask both of you, do you expect the United States to regain some of its former market share, in your case rice and in your case wheat, under a free Iraq? Let us start with you, Mr. King.

Mr. KING. We definitely hope so. It is going to take the GSM credits and other programs to accomplish that, and we had a huge market before and we want it back.

The CHAIRMAN. Mr. Edwards?

Mr. EDWARDS. Yes. We are very confident we can regain those markets. I think as this economy begins to build and the lifestyle improves there, God willing, the need for high-quality wheat will be—well, they will—it will come to us.

The CHAIRMAN. And what about your principal competitors? Do you think any of them have a proximity that will give them an advantage over us, or do you think you can overcome that with your quality and price?

Mr. EDWARDS. I think it is quality and price. I think the major factor will be, as my colleague says, the GSM credit situation and what these other countries have in their toolbox to make deals. I mean, that is where we run up against the State trading enterprise problem that is pervasive in many areas of our trade policy. But have to be able to play on the field, and if other countries—good friends of ours, yes, but competitors nonetheless. If they have other advantages such as credits from their government, for instance, we are going to have to be able to play that game, and I think it is to the advantage of the Iraqi people that they have a relationship with America.

The CHAIRMAN. Are your organizations making any contacts now, as we are in a transitional state of government, that are helping you to build the kind of contacts you need with the new Iraqi government to establish the relationship that will enable you to get a fair treatment in terms of their decisions about contracting for wheat and rice?

Mr. KING. Well, the Rice Council and Rice Federation, yes, we have. We have contacts with the Iraqi Grain Board and we have tried to get them to come to the United States and I think it has been mentioned earlier that they have not been allowed in yet. But that—we definitely have been there and we have contacts and we are ready to—we are going to sell some rice to them.

The CHAIRMAN. Mr. Edwards?

Mr. EDWARDS. Yes, sir. We are in contact constantly with Iraqis. As I said, we met in Jordan not too long back and we have our—we have the contacts. We just have the travel problems in getting them here and us there and all the logistics that come from the upset of war.

The CHAIRMAN. Sure. Well, I hope you will keep this committee informed of your efforts and your progress and let us know if you encounter difficulties. We are very interested in seeing that these markets be reopened to each of your commodities.

The gentleman from Texas, Mr. Stenholm.

Mr. STENHOLM. Continuing along the Chairman's questioning, let us assume for a moment the completion of the DOHA round. Let us assume for a moment that the Europeans have kept their promise to end export subsidies. Let us assume for a moment the Australians and the Canadians gave up their Wheat Boards. Would

U.S. wheat be competitive in the Iraq market and rice without the benefit of credit guarantees?

Mr. KING. We definitely need the credit guarantees. Iraqi people, as it has been mentioned earlier, have financing problems because of the oil pipeline being blown up or whatever.

Mr. STENHOLM. How are we going to make a credible argument that we need credit guarantees when we have asked and sustained and had the others give all of their exports subsidies? How are we going to be credible in our arguments?

Mr. KING. Well, of course, if the other countries have that.

Mr. STENHOLM. The premise of my question is that we have negotiated away all export subsidies and we have negotiated away the Australian and Canadian Wheat Boards.

Mr. EDWARDS. OK.

Mr. STENHOLM. Then would U.S. wheat and rice be competitive in the Iraq market without the benefit of credit guarantees?

Mr. KING. Well, on the rice side, that definitely would just depend on price and what happens in the future. It is a very hard question to answer.

Mr. EDWARDS. Congressman, the export subsidies in our industry in our national organizations are very much in favor of elimination of those everywhere. Only a small part of that would be credit. I prefer to think of this as a credit to a people who are broke. It is not an export subsidy per say in our mind as an industry to set up credit terms that should be repaid.

Mr. STENHOLM. Well, I think it would safe to say that if—and I agree with you on that, the premise basic of both of your testimonies. We have to look at Iraq as a broke nation, and we can't look at the past. We have got to look to this day forward. We realize the United States has lost market share in rice and in wheat. We have lost it under some very questionable practices that are now being investigated by the appropriate U.N. authorities. But that is all in the past. Now what we are looking at is how can we be competitive? How can we regain a share of that market?

If we are going to have credit guarantees, it is reasonable to suppose that other countries will provide credit guarantees, and therefore, we will compete equitably in a level playing field with credit guarantees. That is the premise of my question.

Mr. EDWARDS. And I believe that we should ask our allies, our competitors to be transparent in their dealings. If Iraqi people need credit, we believe they should have credit. We believe as a wheat industry that the allocation of resources to feed people is going to be at least as effective as military expenses.

Mr. STENHOLM. I agree, and this is the subject of the ongoing WTO negotiations now, and that is going to determine the future. But for right now, again, I think we need to use all of the tools in our arsenal to maintain a level playing field for our wheat and our rice producers in this case in the Iraqi market. And I hope that we will have and we will use those resources, and I say will have. It is going to be interesting when we see the mark-up of the agriculture appropriations bill next week as to whether—and how many tools we are going to have left for us in this area.

But that is the short-term. Long-term, I do think that we have to look at Iraq as a new country. Beginning July 1, they are a new country and what has happened in the past, we better—as a free world, all nations had better look at it from the standpoint of doing what we can do to assist them in the area of food, both from a market basis and also from a food aid, and that is always a fine line to walk between the two.

Thank you for your testimony.

The CHAIRMAN. The gentleman from California, Mr. Ose.

Mr. OSE. Thank you, Mr. Chairman.

Gentlemen, I am curious about some—the—I am looking for a little kind of trading process information here. Under the Oil for Food Program—and I understand that the Iraqi government was authorized to sell *X* amount of oil. They would take the revenue, and then the Iraq—Office of the Iraqi—oh, I had it here a minute ago. The OIT would say OK, we are going to go buy food, and they would get tenders for that. Now since the U.N. supervision of the Oil for Food Program ended, have you been able to sell any rice or wheat into Iraq?

Mr. KING. No. We have not sold any rice in Iraq since that time.

Mr. OSE. How about for wheat?

Mr. EDWARDS. Yes. Our sale was made before that program ended, but it was not under the program. But we have made a sale this year.

Mr. OSE. I want to examine that. When you guys did your sale, how did it proceed? Was it Public Law 480 wheat or something else?

Mr. EDWARDS. Let me confer with my expert here. OK. My expert doesn't know, but we will get that information to you.

Mr. OSE. Well, my question is how does it proceed? I mean, does—somebody asks you for a proposal. You respond to it. How do you ensure that you are going to get paid? What I am trying to do is identify where can people who wish to twist the system twist it, to the extent of these commissions and kickbacks and other things that might have occurred previous to your sale?

Mr. EDWARDS. Well, in our experience, that happens when you are dealing with agencies. An Iraqi agency would have more opportunity for leakage or twisting, as you call it, of some of the funds and maybe some kickbacks. When we are able to contact millers and bakers directly and have the direct contact, such as in the 1980's when we brought over millers and grain buyers——

Mr. OSE. Millers and bakers in Iraq?

Mr. EDWARDS. Yes. The Iraqi people.

Mr. OSE. All right.

Mr. EDWARDS. We brought them to the United States, sent them to baking school. There is—we have got a great program at—in Fargo, North Dakota, the Northern Crops Institute, that does some of this education.

Mr. OSE. But that system went by the wayside.

Mr. EDWARDS. Well, we have not been able to have that movement for a decade really.

Mr. OSE. OK.

Mr. EDWARDS. But those are the deals that we like to interact with because then there is not an extra middle man in there. I mean, in the grain industry, there is always middle men as part of it.

Mr. OSE. So essentially since the Oil for Food Program got established, you have had no such direct contact?

Mr. EDWARDS. Essentially, yes.

Mr. OSE. Now in those situations where you had direct contact, you would have the baker or the miller say OK, we need so many tons of wheat delivered at such and such a port by such and such a day?

Mr. EDWARDS. Yes. Our private companies where the seller is out of the United States. Of course, in Australia, and—they use their Wheat Boards. But direct sales are what worked prior to the 1991 Gulf War, and the Iran/Iraq War of course complicated that also. But the direct sales with the users in Iraq is what we have found——

Mr. OSE. The reason, Mr. Chairman, I pursue this line of questioning is that I wanted to make it clear that the system existed previous to the Oil for Food Program for commodity foodstuffs to be delivered without this kind of behavior that seems to have occurred under the Oil for Food Program, and that since the Oil for Food Program got implemented, we have had significant allegations of corruption and what have you. I just wanted to differentiate that the market people, the bakers and the millers, as opposed to the Iraqi government, that there were 2 basically different approaches.

Thank you, Mr. Chairman.

The CHAIRMAN. I thank the gentleman, and I thank both of these witnesses and wish you very well in your efforts to reestablish what was once a very vibrant market for American agricultural exports, and we want to be of assistance to you any way we can. So I reiterate what I said earlier. Keep us informed, and we will work with you every step of the way to make sure that every opportunity is taken to sell food to the Iraqi people. We also call upon you for your support for the continued use of food aid in our international relief efforts, not just in Iraq, but around the world. We find that this is a vitally important program for U.S. foreign policy and for starving people in many corners of the world, and without it, as the representative of USAID said earlier, people will die.

So this is something that American agriculture needs to stand up for, be proud of and to fight for because it is a good cause, and those who would try to subvert it by suggesting that it is something that should be converted into cash, which can disappear into the pockets of wealthy and affluent people, be used to subsidize effectively the agricultural programs of other nations rather than to promote American agriculture since Americans are indeed the most generous people on earth when it comes to helping starving people. We need to make sure that that program is win-win for American agriculture and the needy people of the world. There are always ways we can improve the program, always things we need to do to make sure that we do not have the program interfere with the legitimate markets in countries to encourage the production domestically in countries of various agricultural needs. But those can be paid attention to and still make sure that the food to help starving people is there when it is needed.

I thank you both for your efforts today and your industries efforts in this regard, and with that, I have some magic language that Mr. Ose is definitely interested in because of the fact that we have a need to hold the record open to get answers to some of the very important questions that he asked and documents that he requested from earlier witnesses.

I want to thank all of the witnesses for appearing before the committee today, and the committee will continue to pay close attention to the progress in Iraq's agriculture sector and we will look for increased opportunities for America's farmers and ranchers. The committee will also follow closely the investigations it to the Oil for Food Program in the hopes that all of those responsible for this scandal are held accountable. The work in Iraq will not be easy, but I am confident that a free and democratic Iraq will continue to make progress in agriculture. When the committee visited Iraq last December, we met with many hard working and dedicated Americans and Iraqis who are determined to return this vital industry to the prosperous and stabilized sector it once was.

Without objection, the record of today's hearing will remain open for 10 days to receive additional material and supplemented written responses from witnesses to any question posed by a member of the panel. This hearing on the House Committee on Agriculture is adjourned.

[Whereupon, at 12:12 p.m., the committee was adjourned.]

[Material submitted for inclusion in the record follows:]

STATEMENT OF H. LEE SCHATZ

Mr. Chairman, members of the committee, I appreciate the opportunity to review the work of the U.S. Department of Agriculture in Iraq and assess the outlook for U.S. agricultural exports to that market.

I would like to provide a brief overview of Iraq's agricultural production and highlight the trade opportunities and challenges ahead for U.S. exporters. I also will bring you up-to-date on what USDA and our private sector partners are doing to renew relationships with Iraqi public and private sector partners.

I arrived in Iraq on April 24, 2003, shortly after the fall of Baghdad. As the first USDA employee in Iraq, I served for nine weeks as the senior advisor to the Iraq Ministry of Agriculture for the Office of Reconstruction and Humanitarian Assistance and its successor, the Coalition Provisional Authority (CPA). Since my return to the United States, I have continued to work full time on Iraq issues for USDA.

Ten additional employees from USDA have served in Iraq assisting in reconstruction efforts. USDA has already designated two Foreign Service officers to staff the agriculture office of the new Embassy.

IRAQI GOVERNMENT AGENCIES

USDA has been working primarily with two government agencies—the Ministry of Agriculture (MOA) and the Ministry of Trade (MOT).

MOA focuses on the domestic production of food, fiber, livestock and poultry. This ministry controls and administers land ownership, manages water resources and delivery, may dictate what crops are to be produced, and proposes domestic procurement prices. It imports all production inputs from seeds and fertilizer to fan belts for tractors, veterinary supplies and sprinkler systems. MOA is also responsible for imports of two commodities—corn and protein meal for use in poultry production. The Ministry subsidizes the sales of these inputs to keep food prices low.

MOT imports all food distributed through the Public Distribution System (PDS). The PDS continues to operate, providing every man, woman, and child an adequate food ration meeting goals set earlier in consultation under the Oil For Food program as administered by the United Nations. Even prior to the Oil For Food program, MOT's Grain Board purchased all imported wheat and rice. This ministry contracts for the milling and distribution of wheat through their own and private sector mills to final market points. MOT buys domestic wheat and rice at pre-announced prices. They operate their own truck fleet and grain silos.

IRAQ'S AGRICULTURAL SITUATION

Decades of state intervention in the economy have marginalized private, market-driven initiatives in agriculture and nearly every other sector of the economy. A limited number of large agricultural producers and processors operate in Iraq, but they are, for the most part, technologically still in the 1980's. For the past 20-plus years, Iraq's agriculture and agribusiness effectively have been cut off from innovation. Adoption and use of high-yield varieties, modern herbicides and pesticides, the full range of improved cultural practices (tillage, planting, irrigation, fertilizer use), and new post harvest technologies have largely bypassed Iraq agriculture.

But despite an extended period of mismanagement of its resources, Iraq still has the land, water, and human resources needed for a successful agricultural sector. However, if agriculture is to flourish, it will take time, and require tough decisions on the part of the new leaders in Iraq, including how to budget their money. We estimate that the Ministry of Agriculture, if their new budget allows, will need to import over $1 billion of agricultural inputs annually for Iraq's producers to boost production.

For the next several years, and most likely longer, Iraq will need to rely on imports to meet a large portion of its food and fiber needs, even with substantial gains in production.

PUBLIC SECTOR PURCHASES

To meet current Iraq food needs, the government of Iraq, through MOT, imports and distributes nearly a half million tons of key food items monthly. That requires about 500–600 trucks moving into the country daily. Supplying this Public Distribution System, as it is known, is a huge responsibility. It appears that the Iraqis will continue to fund with their own money this massive public sector purchasing for the near future. In fact, MOT leaders have discussed adding items to the ration to show the Iraqi public that the situation is improving.

MOT has completed its first purchases since resuming responsibility for the PDS, and we are assessing that effort. In the past, Iraq had a 6-month stock of goods in warehouses to avoid shortages, but that luxury is not available today. But MOT, along with the CPA, is working to rebuild at least a three-month buffer stock to better ensure adequate food availability.

We expect that for at least the next year, MOT will remain the major customer for food sales to Iraq. In addition, there may be opportunities for small sales to traders in countries bordering Iraq, who will move those goods into Iraq for the limited private food sector.

But commodities like wheat, rice, and pulses will move almost exclusively through Iraq Government purchases. And the volumes, based on current monthly distribution plans, will be substantial. For the next 12 months, distribution requirements met through imports, will be at or near the following levels:

Wheat 2.6 million tons
Rice 1 million tons
Pulses over 300,000 tons
Ghee and oil over 400,000 tons
Sugar over 600,000 tons
Full fat dry milk over 150,000 tons

Other items included in the PDS are tea, salt, infant formula, weaning cereals, soap, and detergent.

CPA advisors have worked to move Iraqi buying to a more transparent, open system. But the terms of trade used by Iraq in recent years, and likely in the period ahead, are considered quite onerous by U.S. traders. These terms include requirements such as: quality and quantity inspection at the final destination (not the origin of loading) in Iraq; arbitration in Iraq; transportation required on the seller's account to the final warehouse in Iraq; payment 30 days after the presentation of all documents; and no allowances for demurrage. However, we must remember that these same terms found willing suppliers under the Oil for Food (OFF) program.

We all recognize that the previous Iraqi government corrupted the operations of the Oil For Food program. Today, advisors to the Ministries have developed new management teams to change old ways of doing business. Every Ministry now has an independent Office of the Inspector General to audit activities.

ANTICIPATED DEMAND FOR AGRICULTURAL PRODUCTS

We believe that there is much room for growth in the Iraqi market. If we look at the levels of wheat consumption in Iraq we discover that per capita consumption is about 60 percent of the level in Turkey, 70 percent of the level in Iran, and 80 percent of the level in Syria. In the long run, we believe that Iraq will remain a major commercial food market—and a market that will demand high quality from its suppliers.

Iraq has a long history as a commercial market. It was a major market for U.S. agricultural products and our largest market for rice in the late 1980's with annual rice sales in the 400,000 - 500,000 ton range. While the World Food Program (WFP) did have a presence in Iraq, it was not there to provide food aid. WFP had a monitoring (and in the north, distribution) role under the OFF program and earlier this year assisted CPA and the MOT in making commercial purchases. While some isolated food aid for vulnerable groups in Iraq (displaced groups) may be appropriate in the future, in general we do not see a need for U.S. food aid.

USDA's long-term relationship with WFP decision-makers allowed USDA officials to work effectively to ensure that U.S. exporters had an opportunity to compete in the first open and transparent tenders earlier this year. As with any commercial market, U.S. producers had to compete for sales. Our wheat industry demonstrated its competitiveness with sales of 325,000 tons, representing one-third of purchases. U.S. rice, however, faces a different situation. Strong early season U.S. sales this year moved our rice prices sharply above those of our Asian competitors. Although U.S. rice has an image of top quality and Iraq buyers are eager to resume imports, the current premiums for U.S. rice have kept Iraq's purchases focused on Asian supplies.

Status of Export Credit Guarantees

For the remainder of 2004, budgets developed for food imports will allow Iraq to continue commercial cash imports with Iraqi funds.

As you are aware, in 1991 Iraq discontinued all payments against its purchases from the United States under the Commodity Credit Corporation's (CCC) direct sales program, as well as the export credit guarantee programs. That total debt is approximately $1.975 billion; late interest has accrued since then in the amount of $1.92 billion. International discussions, through the Paris Club, have begun on the forgiveness and rescheduling of Iraq's debts. USDA awaits those results and in the interim, USDA is reviewing the possibility of a future guarantee program.

U.S. Government and Private Sector Efforts

Dangerous situations throughout the country have limited our ability to identify and pursue broad capacity building and technical assistance projects. When such work can be undertaken, indications are that the demand for assistance to help Iraq catch-up with 21st century agriculture techniques will be substantial.

For now, USDA has fielded 10 advisors for varying periods of time to work with Iraqis. A major function of USDA's presence in the new Embassy in Baghdad will be to continue to work with the Government of Iraq to identify and provide such assistance to strengthen Iraq's agribusiness sector.

Since late last year, even given the unstable security situation, U.S. private sector trade groups, specifically our market development partners, have been working in Iraq. Their work is limited to training and support from outside the country. U.S. feed grain, soybean, wheat, rice, and pulse cooperators have renewed contacts with Iraqi buyers.

Two of USDA's Market Development Program Cooperators, the U.S. Grains Council and the American Soybean Association, have received USDA funds to help Iraq's broiler and layer producers restart and modernize their industry. A number of technical training seminars for managers of facilities have already been completed. In addition, work is progressing to form a poultry producers association in Iraq. This association would strengthen Iraq private sector involvement in the development of a modern industry.

USDA also supported representatives of the U.S. wheat, rice, and pulse industry to meet with Government of Iraq buyers in Amman, Jordan in February of this year to begin the process of clarifying and modifying contract terms used by Iraq. As a result of these initial meetings, U.S. commodities are once again able to meet Iraq import quality specifications. Contract terms are still an issue, but upcoming meetings with the U.S. industry will continue to work on those outstanding issues.

USDA's Foreign Agricultural Service (FAS) used the Foreign Market Development Program to support the U.S. Grains Council in its proposal to contract an Iraqi specialist to undertake work on behalf of all cooperators desiring to penetrate the Iraqi market. An initial Iraqi contractor has been hired.

Supporting and guiding market promotion activities will be a key focus for the USDA team that will staff the Embassy.

We are currently working to reschedule the first visit to the United States of a team of senior MOT officials. These individuals already know the quality of U.S. agricultural exports, but there is nothing stronger than renewed personal contacts, with our producers, handlers, and exporters to convince Iraq's decision makers that we are ready and able to meet their needs for quality agricultural products.

Mr. Chairman, that concludes my testimony. I would be happy to answer any questions.

ANSWERS TO SUBMITTED QUESTIONS

Is the U.S. Government actively assisting domestic agriculture interested in accessing the Iraqi marketplace? Specifically, I would like to know how growers in my district, who are interested in selling goods to Iraq, can enter the market. In addition, it would be helpful to know who is responsible, if anyone, for providing technical assistance to American agriculture seeking entry in the Iraqi marketplace.

Iraq once was a significant market for U.S. agriculture and we would like to have it back. Before the first Gulf War, Iraq was a one-million-ton market for U.S. wheat and a 500,000-ton market for U.S. rice. The U.S. had total sales of well over $900 million for all food and agricultural products. We want to regain this market. USDA has been focused on getting our products able to compete again from the outset and is actively assisting various U.S. agricultural groups working to re-enter the Iraqi marketplace and re-establish their position.

We currently have several market development activities underway despite the fact that security conditions in Iraq have not allowed traditional Foreign Agricultural Service (FAS) market development cooperators to travel to Iraq.

• We have been working with our rice, wheat, pulses and vegetable oil cooperators on several market development initiatives in Iraq. USDA organized a conference in Amman, Jordan in mid-February with several producer associations to familiarize Iraqi food grain buyers with U.S. quality standards and the structure of the different grain and pulse sectors. The Iraqis and the U.S. sectors represented were pleased to reestablish contact after the long hiatus.

• One of several outcomes of the above seminar was an invitation by the U.S. rice and wheat sectors to the Iraqi Grain Board to visit the United States. The Grain Board is part of the Ministry of Trade. Visa delays and uncertainty related to the transfer of power have postponed the travel from its originally scheduled time in May, but we now are now exploring a delegation visit in September. A two-week program is being planned by FAS, U.S. Wheat Associates, the U.S Rice Federation and the U.S. Rice Producers, with stops in Washington, DC, Kansas and Texas.

• Both of these activities were funded using Market Access Program and Foreign Market Development Program funds.

In the long term, U.S. agriculture should see a continuing market for the key commodities currently distributed by Iraq through the Public Distribution System plus corn and protein meal for the poultry sector in Iraq. However, the private sector will also play a growing role in food trade in the new Iraq. Private businesses are already free to import some food items in Iraq. As their economy grows, Iraq will join other markets in the region and begin to import a broader range of U.S. food products.

FAS Commodity and Marketing Programs (CMP) provides technical assistance for U.S. agricultural exporters seeking entry into foreign markets. CMP manages FAS' targeted market development programs, including the Market Access Program, the Emerging Market Program, Technical Assistance for Specialty Crops, and the Quality Samples Program. Deputy Administrator, Frank Lee, is the head of CMP and may be contacted at (202) 720–4761.

Can you update the committee on the progress that has been made with regard to increasing the buffer stocks since the hearing on June 16th?

In a July 7 phone call with U.S. Government representatives in Baghdad, the level of wheat buffer stocks was reported to be holding at one month's supply of imported wheat.

The domestic wheat harvest is now underway. Procurement by the Ministry of Trade should add up to another two months supply to the national buffer stock by the end of July.

STATEMENT OF WILLIAM GARVELINK

Mr. Chairman and members of the committee it is a pleasure to be here today. I wish to express the apologies of Administrator Natsios who planned to be here today but was called away on an emergency trip to Haiti.

I have his statement for the record and would offer a couple of observations on USAID's role in ensuring that the Iraqi population was fed over the past year-and-a-half.

Prior to the war, as part of the interagency process, humanitarian offices within USAID began preparing for a humanitarian crisis in Iraq and began planning how to guarantee that the Iraqi population continued to receive food supplies through the war and into the post-conflict period. The humanitarian crisis did not happen. But the entire Iraqi population had to be fed. With the full cooperation of the UN World Food Program, the State Department, Coalition Provision Authority and the Civil Affairs Officers of the Coalition Forces, USAID ensured that 26 million Iraqis continued to receive their regular food rations, and they continue to receive them today. Averting a food crisis in Iraqi has been one of the U.S. Government's major successes in Iraq.

By March 2003, we had secured the agreement of the World Food Program to assume the overall management of the Oil for Food program, using Iraqi's existing Public Distribution System. For years in Iraqi, each Iraqi citizen has received a monthly food ration. These rations have been distributed through a network of almost 45,000 food and flour agents. Rather than create a new system, the World Food Program revived this network and used it to feed the Iraqi population, and it is being used today.

Managing this national food ration system has been an enormous task. It requires almost 500,000 metric tons of commodities to be distributed throughout the country monthly. A fleet of thousands of private and public sector trucks make daily trips in and out of Iraq, hauling food overland to more than 400 warehouses and silos around the country. 45,000 food agents pick up these commodities for the final distribution to Iraqi families.

As the war began in March, the United States assumed that Iraqi families had a 2–3 month food supply. To ensure that there would be food available after their supplies ran out, possibly in June, USAID took several actions. USAID provided $45 million to the World Food Program to fund its mobilization of staff and initial trucking and fuel contracts. USAID provided to the World Food Program $200 million in cash and $181 million in P.L. 480 Title II and Emerson Trust food commodities. These actions total $426 million. On June 1, 2003, food distributions began through the Iraqi Public Distribution System and have continued ever since.

Through the intervening months, the World Food Program, Coalition Provisional Authority and USAID have worked with the Iraqi Ministry of Trade to train staff and prepare the Ministry for the time when it would assume full control of the Public Distribution System. On June 30 the Ministry will take over the system and the last commodities procured by the World Food Program will arrive by September 2004.

We fully support the handover of the Public Distribution System to the Interim Iraqi Government, but we remain concerned that insecurity and management challenges might lead to disruptions in the food pipeline. Delays in food deliveries not only threaten the vulnerable, they could also create further instability among the population. We will remain vigilant in the coming months and be prepared to assist the new Iraqi government should it become necessary.

I cannot end my comments without noting that 2004 marks the 50th anniversary of the P.L. 480 Food for Peace Program. Over the past 50 years, the U.S. Government has contributed more than $50 billion to finance more than 367 million metric tons of food to over 150 countries. More than 3.3 billion people worldwide have benefited from this program. It is a credit to the dedication and vigilance of this committee and its Members that the United States has been able to share its bounty with so many of the world's needy. We are all justly very proud of this program.

Thank you.

STATEMENT OF ANDREW S. NATSIOS

Mr. Chairman and members of the committee, I am pleased to have the opportunity to discuss with you today the U.S. Government's successful efforts to avert a humanitarian crisis in the aftermath of the war in Iraq.

In late 2002, Iraq's Hussein regime continued to defy the United Nations Security Council resolutions, ignoring warnings from the international community about the "grave consequences" to follow. As the prospect for war became imminent, the U.S. Agency for International Development (USAID) began to plan for a possible humanitarian crisis in Iraq. The situation in Iraq was made infinitely more complex because of that country's dependence on the United Nations Oil for Food (OFF) program and a public food ration system that fed the entire population of 26 million people.

USAID was asked to join an interagency working group. At their direction, USAID initiated discussions with the United Nations World Food Program (WFP) to determine our response options. We agreed that disruption of the public food distribution system, or PDS, would require major interventions by the United States to ensure continued food distributions to the Iraqi people and to protect Iraq's most vulnerable groups. We prepared a three-fold strategy should conflict erupt: restart the PDS as soon as possible; feed the entire population of 26 million, and be prepared to feed several million internally displaced persons (IDPs) and refugees if there were a humanitarian crisis.

Through extensive planning, sufficient U.S. resources, and the hard work of USAID, other U.S. Government agencies, the World Food Program, the Iraqi Ministry of Trade, and the U.S. military's civil affairs personnel, we were able to reestablish the PDS in less than 30 days following major combat operations. On June 1, 2003, the food pipeline was open in all 18 governorates. This averted a humanitarian food crisis and maintains an acceptable level of PDS food supplies throughout the country today.

Managing this unique food distribution system was an enormous effort. Under typical operations, the PDS provides almost 500,000 tons of food to Iraq every month. The monthly ration includes flour, rice, cooking oil, lentils or beans, sugar, tea, salt, powdered milk, detergent, soap and infant formula. Last year, Oil for Food resources were used to purchase these commodities. Today, funds from the Iraqi operating budget fund the purchase of commodities that are shipped to Iraq and other regional port facilities. A fleet of thousands of private and public sector trucks then make daily trips in and out of Iraq, hauling the food overland to more than 400 warehouses and silos throughout the county. Finally, local food and flour agents travel to the warehouses, receive their assigned shipments, and return to their communities for final distributions. Each person in Iraq is qualified for a food ration and families are assigned ration cards that allow them to receive monthly rations from the agents who distribute the commodities from storefronts or private residences. A centralized database maintained by the Ministry of Trade provided population data and monitored births, deaths and relocations.

One of our early concerns was that this extensive system would become disrupted during the war and that food suppliers would refuse to fulfill their food contracts due to security issues. We also were unsure of the food stock supplies in Iraq prior to the war. The U.S. Government estimated as much as three months supply of food, however, we also knew that the regime had distributed double rations for several months in 2002 and early 2003. We were unsure how long food supplies would last, and we needed to identify and secure resources for additional food supplies should they be needed.

On March 28, 2003, WFP issued a worldwide emergency appeal for $1.3 billion to fund a six-month operation to stand up the PDS and keep the food moving. To help ensure an initial food supply line, USAID contributed $45,000,000 in cash to support WFP's preparation and mobilization costs, allowing WFP to position food and personnel for the tasks ahead. In addition, we contributed $200,000,000 in cash and almost $181,000,000 worth of Title II and Emerson Trust food commodities to the WFP operation. In dollar terms, this became USAID's second largest food assistance program in fiscal year 2003, totaling $426 million, and provided a total of 575,320 metric tons of food to the assistance effort.

In addition to food and cash, the USAID Disaster Assistance Response Team, or DART, helped backstop other critical WFP operations. Food experts were deployed with the DART and assigned to the key Iraqi cities of Basrah, Baghdad, Hillah, and Erbil, as well as in Kuwait City and Cyprus. Throughout the spring and summer of 2003, the DART food team assisted WFP, the Iraqi Ministry of Trade, and the Coalition Forces with numerous food distribution activities. This included securing warehouse stocks, assuring records and data were secure, liaising with civilian-military operations, prioritizing humanitarian efforts, assisting WFP Iraqi staff with logistics and obtaining military de-mining and other technical assistance. In September of 2003, we were confident that a food crisis had been averted, and the DART returned to the United States.

AFTER THE CONFLICT

Having averted a food crisis immediately following the conflict, USAID continued to focus and assist with the management and distribution of food rations for all Iraqi citizens. We soon realized that the Iraqis were capable of managing much of the PDS tasks, but needed some training and equipment. Again, in partnership with WFP, the newly organized Coalition Provisional Authority (CPA) and the military, we began to take immediate actions to support the Iraqi food administrators so they could effectively take the reigns of the PDS.

Some of these actions include assisting CPA and the Iraqi government with OFF handover tasks, providing technical expertise to identify thousands of former OFF-funded projects and resources for the Iraqi people, including food contracts for the PDS. We also played a key role in establishing the Baghdad Coordination Center, an essential technical service center to be operated by the Iraqis that will help assure proper and transparent management of the former OFF activities. We continue to contribute counterpart personnel to provide logistics expertise, supplier service support, contract reference and processing support, payment services, UN liaison and other essential services to help ensure that the food supply continues to flow. We are providing food specialists in Baghdad to serve as technical advisors to the Iraqis, helping arrange procurement training sessions with WFP in Rome and communications training in Amman, Jordan. Finally, we provided ministries with additional computer equipment to help improve communications and computing capabilities.

THE FUTURE

Iraq's PDS continues to provide food to all Iraqi citizens. Though we realize that independent market forces also contribute to food supplies in Iraq, we have not reduced our concern for Iraq's most vulnerable groups nor our determination to assist the Iraqi's with their PDS operations, if needed. We also recognize the PDS is extremely costly to the government and inherently inefficient. But as long as Iraqis are dependent on food rations, we will remain attentive to Iraq's food supply and involved.

In addition to our assistance with the PDS, we have begun other food security related activities to reduce risk for the most in need. In April of this year, in collaboration with the Ministry of Agriculture and the CPA, we prepared an Agricultural Transition Plan designed to restore agricultural production and create jobs to bring about a transition to a market-led agriculture sector. We plan to support the Ministry's efforts in policy and economic analysis, enforcement of a regulatory framework to protect human health and the environment, and with applied research and farmer extension. If rehabilitated and managed properly, we believe that the agricultural sector in Iraq will have the physical and human resources that can significantly contribute to economic recovery in generating employment and income and in improving food security. Increased food security and a revived agricultural sector will allow Iraq to wean its population from the PDS, opening the way for food sufficiency and free market enterprise.

It should be underscored that until these and other infrastructure, institutional building and economic improvement efforts begin to impact Iraqi society, many Iraqis will remain dependent on the PDS as the main source of food. Although we fully support the handover of PDS responsibilities to the interim government on June 30, we remain fearful that interruptions in the PDS food pipeline, resulting in food shortages. Delayed food deliveries not only threaten the vulnerable, they could create further instability among the population.

We applaud the hard work of the talented and dedicated people oft the CPA, WFP, and the countless others who have made this effort one of the U.S. Government's most successful in Iraq. Yet we by no means believe the task is finished. Several key technical improvements are still necessary for the successful continuation of the PDS aftr the CPA closes its doors. A critical concern is the tenuous nature of the food supply pipeline. We must remain vigilant in the upcoming months to assist the new Iraqi government should it be necessary.

50 YEARS OF U.S. FOOD AID

Before I conclude, let me add that this year marks the 50th anniversary of the Food for Peace program. Over the past 50 years, the U.S. Government has contributed more than $50 billion to finance more than 367 million metric tons of food to over 150 food insecure countries. More than 3.3 billion people worldwide have been recipients of U.S. food assistance. In addition to the successful effort in Iraq, other countries worldwide have benefited from this long-standing U.S. assistance program.

In much of Asia and Latin America, where famine has taken millions of lives, basic food security has been established and sustainable development has become a reality. In many of these places, our food aid investments have played an important role in helping to bring this about. Japan, South Korea, Mexico, Argentina, Morocco, Chile, and even a number of European countries were recipients of early P.L. 480 programs. Many of these graduated countries have gone on to become major food donors.

Some of the most notable achievements occurred in the aversion of widespread famine in Ethiopia; the feeding of four and a half million people in southern Africa; the prevention of a humanitarian food crisis in Iraq and the successful response to Afghanistan's drought—which included the shipment of nearly 400,000 metric tons of food to that country, benefiting 10 million people between 1999 and 2002. The World Food Program played a key role in these successes.

The year 2003 saw one of the largest amounts of food distributed through Title II in its 50-year history—and just slightly below what was distributed in 1985, the year of the terrible Ethiopian famine. In fact, if you add the 401,000 metric tons we received from the Bill Emerson Humanitarian Trust, last year was the biggest year since P.L. 480 was passed.

Despite the investments and the progress made over the past 50 years, nearly 840 million people are still food insecure. And though critical for addressing short-term food needs and saving lives, USAID recognizes that simply feeding people from one day to the next is not going to end hunger.

Our food aid will obviously continue to focus on people's short-term food and nutritional needs. But we have to invest in longer-term solutions, as well. It does not make sense to spend hundreds of millions of dollars feeding people in a country like Ethiopia and only provide a tiny fraction of that to help them improve their agriculture.

Consequently, USAID is working toward implementation of both short and long-term interventions which link agricultural development, trade and food aid in order to promote food security. The United States is committed to supporting a variety of proven and innovative programs to address hunger, and USAID's Office of Food for Peace is making a dramatic shift in its approach to addressing food needs.

The American people can be proud of the many accomplishments of the Food for Peace program over the years. I look forward to working with you on both responding to emergencies and to improving food security abroad.

GAO
Accountability·Integrity·Reliability

Highlights

Highlights of GAO-04-880T, a testimony before the House Committee on Agriculture

Why GAO Did This Study

The Oil for Food program was established by the United Nations and Iraq in 1996 to address concerns about the humanitarian situation after international sanctions were imposed in 1990. The program allowed the Iraqi government to use the proceeds of its oil sales to pay for food, medicine, and infrastructure maintenance. The program appears to have helped the Iraqi people. From 1996 through 2001, the average daily food intake increased from 1,300 to 2,300 calories. From 1997-2002, Iraq sold more than $67 billion of oil through the program and issued $38 billion in letters of credit to purchase commodities. However, over the years numerous allegations have surfaced concerning potential fraud and program mismanagement.

GAO (1) reports on its estimates of the illegal revenue acquired by the former Iraqi regime in violation of U.N. sanctions, (2) provides observations on program administration, and (3) describes the current and future challenges in achieving food security.

www.gao.gov/cgi-bin/getrpt?GAO-04-880T.

To view the full product, including the scope and methodology, click on the link above. For more information, contact Joseph Christoff at (202) 512-8979 or christoffj@gao.gov.

Wednesday, June 16, 2004

UNITED NATIONS

Observations on the Oil for Food Program and Iraq's Food Security

What GAO Found

GAO estimates that from 1997- 2002, the former Iraqi regime acquired $10.1 billion in illegal revenues, including $5.7 billion in oil smuggled out of Iraq and $4.4 billion through surcharges on oil sales and illicit commissions from suppliers exporting goods to Iraq through the Oil for Food program. This estimate includes oil revenue and contract amounts for 2002, updated letters of credit from prior years, and newer estimates of illicit commissions from commodity suppliers.

The United Nations, through the Office of the Iraq Program (OIP) and the Security Council's Iraq sanctions committee, were both responsible for overseeing the Oil for Food Program. However, the Security Council allowed the Iraq government, as a sovereign entity, to negotiate contracts directly with purchasers of Iraqi oil and suppliers of commodities. This structure was an important factor in allowing Iraq to levy illegal surcharges and commissions. OIP was responsible for examining Iraqi contracts for price and value, but it is unclear how it performed this function. The sanctions committee was responsible for monitoring oil smuggling, screening contracts for items that could have military uses, and approving oil and commodity contracts. The sanctions committee took action to stop illegal oil surcharges, but it is unclear what actions it took on contract commissions. U.N. external audit reports contained no findings of program fraud. Summaries of internal audit reports pointed to some concerns regarding procurement, coordination, monitoring, and oversight and concluded that OIP had generally responded to audit recommendations. Ongoing investigations of the Oil for Food program may wish to further examine how the structure of the program enabled the Iraqi government to obtain illegal revenues, the role of member states in monitoring and enforcing the sanctions, actions taken to reduce oil smuggling, and the responsibilities and procedures for assessing price reasonableness in commodity contracts.

Evolving policy and implementation decisions on the food distribution system and the worsening security situation have affected the movement of food commodities within Iraq. As a result, as of June 2004, food warehouse stocks are low and Iraq has less than a month's supply of essential food items, according to U.S. and World Food Program officials. In addition to these current food security challenges, the new government will have to balance the need to reform a costly food subsidy program with the need to maintain food stability and protect the poorest populations. Also, inadequate oversight and corruption in the Oil for Food program raise concerns about the Iraqi government's ability to manage the food distribution system and absorb $32 billion in expected donor funds for reconstruction. The coalition authority has taken steps, such as appointing inspectors general, to build internal controls and accountability measures in Iraq's ministries.

Mr. Chairman and Members of the Committee:

I am pleased to be here today to discuss GAO's review of the United Nations (U.N.) Oil for Food program and Iraq's food security.

In 1996, the United Nations and Iraq established the Oil for Food program to address growing concerns about the humanitarian situation after international sanctions were imposed in 1990. The program's intent was to allow the Iraqi government to use the proceeds of its oil sales to pay for food, medicine, and infrastructure maintenance and, at the same time, prevent the regime from obtaining goods for military purposes. From 1997 through 2002, Iraq sold more than $67 billion in oil through the program and issued $38 billion in letters of credit to purchase commodities.[1]

Today, we will (1) report on our estimates of the illegal revenue acquired by the former Iraqi regime in violation of U.N. sanctions, (2) provide our observations on the administration of the program, and (3) describe the current and future challenges in achieving food security.

To address these objectives, we reviewed documents and statements from (1) the United Nations on its management and oversight responsibilities for the Oil for Food program; (2) the Coalition Provisional Authority (CPA), the Departments of Defense and State, and the United Nations and its World Food Program (WFP) on the current implementation of the program and challenges in ensuring a continued food supply; and (3) from the World Bank and Iraq's 2004 budget regarding the effect of food subsidies on the Iraqi economy. We also reviewed all external audits to determine the use of Oil for Food funds prior to the transfer of the program to the CPA in November 2003. We did not have full access to the U.N. internal audits of the Oil for Food program, but we reviewed the summaries of 7 annual internal audits from 1996 to 2003 and had access to one report made publicly available in May 2004. We met with U.N. officials following the transfer of the program to the CPA and with numerous U.S. officials representing the CPA, the Departments of Defense and State, and the U.S. Agency for International Development, to examine current program management, the status of the food distribution system, and current and future food security challenges.

[1] All references to Oil for Food estimates are in 2003 constant U.S. dollars.

We conducted our review from November 2003 through June 2004 in accordance with generally accepted government auditing standards.

Summary

- From 1997 through 2002, we estimate that the former Iraqi regime acquired $10.1 billion in illegal revenues—$5.7 billion in oil smuggled out of Iraq and $4.4 billion in surcharges on oil sales and illicit charges from suppliers exporting goods to Iraq through the Oil for Food program. This estimate is higher than our May 2002 estimate of $6.6 billion because it includes (1) oil revenue and contract amounts for 2002, (2) updated letters of credit from prior years, and (3) newer estimates of illicit commissions from commodity suppliers.
- The United Nations, through the Office of the Iraq Program (OIP) and the Security Council's Iraq sanctions committee, were responsible for overseeing the Oil for Food Program. However, the Security Council allowed the Iraqi government, as a sovereign entity, to negotiate contracts directly with purchasers of Iraqi oil and suppliers of commodities. This structure was an important factor in enabling Iraq to levy illegal surcharges and commissions. OIP was responsible for examining Iraqi contracts for price and value, but it is unclear how it performed this function. The sanctions committee was responsible for monitoring oil smuggling, screening contracts for items that could have military uses, and approving oil and commodity contracts. The sanctions committee took action to stop illegal surcharges on oil, but it is unclear what actions it took on the commissions on commodity contracts. U.N. external audit reports contained no findings of program fraud. Summaries of internal audit reports provided to GAO pointed to some operational concerns in procurement, coordination, monitoring, and oversight. Ongoing investigations of the Oil for Food program may wish to further examine how the structure of the program enabled the Iraqi government to obtain illegal revenues, the role of member states in monitoring and enforcing the sanctions, actions taken to reduce oil smuggling, and the responsibilities and procedures for assessing price reasonableness in commodity contracts.
- Evolving policy and implementation decisions on the food distribution system and the worsening security situation have affected the movement of food commodities within Iraq. As a result, according to U.S. and WFP officials, food warehouse stocks are low and the country has less than a month's supply of essential food items, such as staple grains. In addition to these current food security challenges, the new government will have to balance the need to reform a costly food subsidy program with the need to maintain food stability and protect

the poorest populations. Also, inadequate oversight and corruption in the Oil for Food program raise concerns about the Iraqi government's ability to manage the food distribution system and absorb $32 billion in expected donor funds for reconstruction. The CPA has taken steps, such as appointing inspectors general, to build internal controls and accountability measures in Iraq's ministries.

Background

In August 1990, Iraq invaded Kuwait, and the United Nations imposed sanctions against Iraq. Security Council resolution 661 of 1990 prohibited all nations from buying and selling Iraqi commodities, except for food and medicine. Security Council resolution 661 also prohibited all nations from exporting weapons or military equipment to Iraq and established a sanctions committee to monitor compliance and progress in implementing the sanctions. The members of the sanctions committee were members of the Security Council. Subsequent Security Council resolutions specifically prohibited nations from exporting to Iraq items that could be used to build chemical, biological, or nuclear weapons. In 1991, the Security Council offered to let Iraq sell oil under a U.N. program to meet its peoples' basic needs. The Iraqi government rejected the offer, and over the next 5 years, the United Nations reported food shortages and a general deterioration in social services.

In December 1996, the United Nations and Iraq agreed on the Oil for Food program, which permitted Iraq to sell up to $1 billion worth of oil every 90 days to pay for food, medicine, and humanitarian goods. Subsequent U.N. resolutions increased the amount of oil that could be sold and expanded the humanitarian goods that could be imported. In 1999, the Security Council removed all restrictions on the amount of oil Iraq could sell to purchase civilian goods. The United Nations and the Security Council monitored and screened contracts that the Iraqi government signed with commodity suppliers and oil purchasers, and Iraq's oil revenue was placed in a U.N.-controlled escrow account. In May 2003, U.N. resolution 1483 requested the U.N. Secretary General to transfer the Oil for Food program to the CPA by November 2003. (Appendix I contains a detailed chronology of Oil for Food program and sanctions events.) The United Nations allocated 59 percent of the oil revenue for the 15 central and southern governorates, which were controlled by the central government; 13 percent for the 3 northern Kurdish governorates; 25 percent for a war reparations fund for victims of the Iraq invasion of Kuwait in 1990; and 3 percent for U.N. administrative costs, including the costs of weapons inspectors.

From 1997 to 2003, the Oil for Food program was responsible for more than $67 billion of Iraq's oil revenue. Through a large portion of this revenue, the United Nations provided food, medicine, and services to 24 million people and helped the Iraqi government supply goods to 24 economic sectors. Despite concerns that sanctions may have worsened the humanitarian situation, the Oil for Food program appears to have helped the Iraqi people. According to the United Nations, the average daily food intake increased from around 1,275 calories per person per day in 1996 to about 2,229 calories at the end of 2001. Malnutrition rates for children under 5 fell by more than half. In February 2002, the United Nations reported that the Oil for Food program had considerable success in several sectors such as agriculture, food, health, and nutrition by arresting the decline in living conditions and improving the nutritional status of the average Iraqi citizen.

Since 1997, Iraq has imported almost 2.7 million metric tons of wheat annually. During the 1980s, Australia was Iraq's primary wheat supplier with 38 percent of the market, and the United States was the second major supplier at 29 percent. By 1989, Iraq was the twelfth largest market for U.S. agricultural exports, including rice. Since 1997, however, Australia has dominated Iraq's Oil for Food wheat trade with a 73 percent market share, and Vietnam has become a major supplier of rice to Iraq. The U.S. market share for wheat dropped to 6 percent during that time. U.S. wheat exports during the sanctions only occurred in 1997 and 1998.

Former Iraqi Regime Acquired an Estimated $10.1 Billion in Illicit Revenue

We estimate that, from 1997 through 2002, the former Iraqi regime acquired $10.1 billion in illegal revenues—$5.7 billion through oil smuggled out of Iraq and $4.4 billion through surcharges against oil sales and illicit commissions from commodity suppliers. This estimate is higher than the $6.6 billion in illegal revenues we reported in May 2002.[2] We updated our estimate to include (1) oil revenue and contract amounts for 2002, (2) updated letters of credit from prior years, and (3) newer estimates of illicit commissions from commodity suppliers. Appendix II describes our methodology for determining illegal revenues gained by the former Iraqi regime.

[2]U.S. General Accounting Office, *Weapons of Mass Destruction: U.N. Confronts Significant Challenges in implementing Sanctions Against Iraq*, GAO-02-625 (Washington, D.C.: May 23, 2002).

Oil was smuggled out through several routes, according to U.S. government officials and oil industry experts. Oil entered Syria by pipeline, crossed the borders of Jordan and Turkey by truck, and was smuggled through the Persian Gulf by ship. Jordan maintained trade protocols with Iraq that allowed it to purchase heavily discounted oil in exchange for up to $300 million in Jordanian goods. Syria received up to 200,000 barrels of Iraqi oil a day in violation of the sanctions. Oil smuggling also occurred through Turkey and Iran.

In addition to revenues from oil smuggling, the Iraqi government levied surcharges against oil purchasers and commissions against commodity suppliers participating in the Oil for Food program. According to some Security Council members, the surcharge was up to 50 cents per barrel of oil and the commission was 5 to 15 percent of the commodity contract.

In our 2002 report, we estimated that the Iraqi regime received a 5-percent illicit commission on commodity contracts. However, a September 2003 Department of Defense review found that at least 48 percent of 759 Oil for Food contracts that it reviewed were potentially overpriced by an average of 21 percent.[3] Food commodity contracts were the most consistently overpriced, with potential overpricing identified in 87 percent of the contracts by an average of 22 percent. The review also found that the use of middlemen companies potentially increased contract prices by 20 percent or more. Defense officials found 5 contracts that included "after-sales service charges" of between 10 and 20 percent.

In addition, interviews by U.S. investigators with high-ranking Iraqi regime officials, including the former oil and finance ministers, confirmed that the former regime received a 10-percent commission from commodity suppliers. According to the former oil minister, the regime instituted a fixed 10-percent commission in early 2001 to address a prior "compliance" problem with junior officials. These junior officials had been reporting lower commissions than what they had negotiated with suppliers and pocketing the difference.

[3]The Defense Contract Audit Agency and the Defense Contract Management Agency, *Report on the Pricing Evaluation of Contracts Awarded under the Iraq Oil for Food Program* (Washington, D.C.: Sept. 12, 2003).

United Nations and Security Council Had Responsibility for Oversight of Program, but Iraq Contracted Directly with Purchasers and Suppliers

Both OIP, as an office within the U.N. Secretariat, and the Security Council's sanctions committee were responsible for overseeing the Oil for Food Program. However, the Iraqi government negotiated contracts directly with purchasers of Iraqi oil and suppliers of commodities. While OIP was to examine each contract for price and value, it is unclear how it performed this function. The sanctions committee was responsible for monitoring oil smuggling, screening contracts for items that could have military uses, and approving oil and commodity contracts. The sanctions committee responded to illegal surcharges on oil purchases, but it is unclear what actions it took to respond to commissions on commodity contracts. Ongoing investigations of the Oil for Food program may wish to consider further examining how the structure of the program enabled the Iraqi government to obtain illegal revenues, the role of member states in monitoring and enforcing the sanctions, actions taken to reduce oil smuggling, and the responsibilities and procedures for assessing price reasonableness in commodity contracts.

Iraq Negotiated Directly with Oil Purchasers and Suppliers

U.N. Security Council resolutions and procedures recognized the sovereignty of Iraq and gave the Iraqi government authority to negotiate contracts and decide on contractors. Security Council resolution 986 of 1995 authorized states to import petroleum products from Iraq, subject to the Iraqi government's endorsement of transactions. Resolution 986 also stated that each export of goods would be at the request of the government of Iraq. Security Council procedures for implementing resolution 986 further stated that the Iraqi government or the United Nations Inter-Agency Humanitarian Program would contract directly with suppliers and conclude the appropriate contractual arrangements. Iraqi control over contract negotiations was an important factor in allowing Iraq to levy illegal surcharges and commissions.

When the United Nations first proposed the Oil for Food program in 1991, it recognized this vulnerability. At that time, the Secretary General proposed that the United Nations, an independent agent, or the government of Iraq be given the responsibility to negotiate contracts with oil purchasers and commodity suppliers. The Secretary General concluded that it would be highly unusual or impractical for the United Nations or an independent agent to trade Iraq's oil or purchase commodities. He recommended that Iraq negotiate the contracts and select the contractors. However, he stated that the United Nations and Security Council would have to ensure that Iraq's contracting did not circumvent the sanctions and was not fraudulent. The Security Council further proposed that U.N.

agents review contracts and compliance at Iraq's oil ministry, but Iraq refused these conditions.

OIP Was Responsible for Key Oversight Aspects of the Program

OIP administered the Oil for Food program from December 1996 to November 2003. Under Security Council resolution 986 of 1995 and a memorandum of understanding between the United Nations and the Iraqi government, OIP monitored the sale of Iraq's oil, monitoring Iraq's purchase of commodities and the delivery of goods, and accounting for the program's finances. The United Nations received 3 percent of Iraq's oil export proceeds for its administrative and operational costs, which included the cost of U.N. weapons inspections.

The sanctions committee's procedures for implementing resolution 986 stated that independent U.N. inspection agents were responsible for monitoring the quality and quantity of the oil shipped. The agents were authorized to stop shipments if they found irregularities. OIP hired a private firm to monitor Iraqi oil sales at exit points. However, the monitoring measures contained weaknesses. According to U.N. reports and a statement from the monitoring firm, the major offshore terminal at Mina al-Basra[4] did not have a meter to measure the oil pumped nor could onshore storage capacity be measured. Therefore, the U.N. monitors could not confirm the volume of oil loaded onto vessels. Also, in 2001, the oil tanker *Essex* took a large quantity of unauthorized oil from the platform when the monitors were off duty. In December 2001, the Security Council required OIP to improve the monitoring at the offshore terminal. As part of its strategy to repair Iraq's oil infrastructure, the CPA plans to install reliable metering at Mina al-Basra and other terminals, but no contracts have been let.

OIP also was responsible for monitoring Iraq's purchase of commodities and the delivery of goods. Security Council resolution 986, paragraph 8a(ii) required Iraq to submit a plan, approved by the Secretary General, to ensure equitable distribution of Iraq's commodity purchases. The initial distribution plans focused on food and medicines while subsequent plans were expansive and covered 24 economic sectors, including electricity, oil, and telecommunications.

[4]Previously called Mina al-Bakar.

The sanctions committee's procedures for implementing Security Council resolution 986 stated that experts in the Secretariat were to examine each proposed Iraqi commodity contract, in particular the details of price and value, and to determine whether the contract items were on the distribution plan. OIP officials told the Defense Contract Audit Agency they performed very limited, if any, pricing review. They stated that no U.N. resolution tasked them with assessing the price reasonableness of the contracts and no contracts were rejected solely on the basis of price. However, OIP officials stated that, in a number of instances, they reported to the sanctions committee that commodity prices appeared high, but the committee did not cite pricing as a reason to place holds on the contracts. For example, in October 2001, OIP experts reported to the sanctions committee that the prices in a proposed contract between Iraq and the Al-Wasel and Babel Trading Company appeared high. However, the sanctions committee reviewed the data and approved the contract. Subsequently, the Treasury Department identified this company as a front company for the former regime in April 2004. The United Nations also required all countries to freeze the assets of this company and transfer them to the Development Fund for Iraq in accordance with Security Council resolution 1483.[5]

The sanctions committee's procedures for implementing resolution 986 stated that independent inspection agents will confirm the arrival of supplies in Iraq. OIP deployed about 78 U.N. contract monitors to verify shipments and authenticate the supplies for payment. OIP employees were able to visually inspect 7 to 10 percent of the approved deliveries.

Audits Identified Some Operational Concerns but No Fraud

Security Council resolution 986 also requested the Secretary General to establish an escrow account for the Oil for Food Program and to appoint independent and certified public accountants to audit the account. The Secretary General established an escrow account at BNP Paribas for the deposit of Iraqi oil revenues and the issue of letters of credit to suppliers with approved contracts. The U.N. Board of Audit, a body of external public auditors, audited the account.[6] The external audits focused on

[5]U.N. Security Council Res. 1483 (May 22, 2003). Paragraph 19 states that a Security Council committee will identify individuals and entities whose financial assets should be transferred to the Development Fund for Iraq.

[6]The U.N. Board of Auditors is comprised of the Auditors General of three member countries and their staff. Board members are appointed by the General Assembly for 6-year terms and one member rotates every 2 years. During the period of the Oil for Food program (1996-2003), France, Ghana, India, the Philippines, South Africa, and the United Kingdom served on the Board of Auditors.

management issues related to the Oil for Food program and the financial condition of the Iraq account. U.N. auditors generally concluded that the Iraq account was fairly presented in accordance with U.N. financial standards. The reports stated that OIP was generally responsive to external audit recommendations. The external audits determined that oil prices were mostly in accordance with the fair market value of oil products to be shipped and checked to confirm that pricing was properly and consistently applied. They also determined that humanitarian and essential services supplies procured with oil funds generally met contract terms with some exceptions. U.N. external audit reports contained no findings of fraud during the program.

The U.N. Office of Internal Oversight Services (OIOS) conducted internal audits of the Oil for Food program and reported the results to OIP's executive director. OIOS officials stated that they have completed 55 audits and have 4 ongoing audits of the Oil for Food program. Overall, OIOS reported that OIP had made satisfactory progress in implementing most of its recommendations. We did not have access to individual OIOS audit reports except for an April 2003 report made publicly available in May 2004 that assessed the activities of the company contracted by the United Nations to authenticate goods coming into Iraq. It found that the contractor did not perform all required duties and did not adequately monitor goods coming into the northern areas of Iraq. We also reviewed 7 brief summaries of OIOS reports covering the Oil for Food program from July 1, 1996, through June 30, 2003. These summaries identified a variety of operational concerns involving procurement, inflated pricing and inventory controls, coordination, monitoring, and oversight. In one case, OIOS cited purchase prices for winter items for displaced persons in northern Iraq that were on average 61 percent higher than local vendor quotes obtained by OIOS. In another case, an OIOS review found that there was only limited coordination of program planning and insufficient review and independent assessment of project implementation activities.

The Sanctions Committee Had a Key Role in Enforcing Sanctions and Approving Contracts

The sanctions committee was responsible for three key elements of the Oil for Food Program: (1) monitoring implementation of the sanctions, (2) screening contracts to prevent the purchase of items that could have military uses, and (3) approving Iraq's oil and commodity contracts.

U.N. Security Council resolution 661 of 1990 directed all states to prevent Iraq from exporting products, including petroleum, into their territories. Paragraph 6 of resolution 661 established a sanctions committee to report to the Security Council on states' compliance with the sanctions and to recommend actions regarding effective implementation. As early as June 1996, the Maritime Interception Force, a naval force of coalition partners including the United States and Great Britain, informed the sanctions committee that oil was being smuggled out of Iraq through Iranian territorial waters. In December 1996, Iran acknowledged the smuggling and reported that it had taken action. In October 1997, the sanctions committee was again informed about smuggling through Iranian waters. According to multiple sources, oil smuggling also occurred through Jordan, Turkey, Syria, and the Gulf. Smuggling was a major source of illicit revenue for the former Iraqi regime through 2002.

A primary function of the members of the sanctions committee was to review and approve contracts for items that could be used for military purposes. The United States conducted the most thorough review; about 60 U.S. government technical experts assessed each item in a contract to determine its potential military application. According to U.N. Secretariat data in 2002, the United States was responsible for about 90 percent of the holds placed on goods to be exported to Iraq. As of April 2002, about $5.1 billion worth of goods were being held for shipment to Iraq. According to OIP, no contracts were held solely on the basis of price.

Under Security Council resolution 986 of 1995 and its implementing procedures, the sanctions committee was responsible for approving Iraq's oil contracts, particularly to ensure that the contract price was fair, and for approving Iraq's commodity contracts. The U.N.'s oil overseers reported in November 2000 that the oil prices proposed by Iraq appeared low and did not reflect the fair market value.[7] According to a senior OIP official, the independent oil overseers also reported in December 2000 that purchasers of Iraqi oil had been asked to pay surcharges. In March 2001, the United States informed the sanctions committee about allegations that Iraqi government officials were receiving illegal surcharges on oil contracts and illicit commissions on commodity contracts. The sanctions committee attempted to address these allegations by implementing retroactive pricing for oil contracts in 2001.[8]

[7]The sanctions committee received reports from the independent oil experts appointed by the Secretary General to determine whether there was fraud or deception in the oil contracting process.

[8]Under retroactive pricing, the Security Council did not approve a price per barrel until the oil was delivered to the refinery. The Iraq government signed contracts with suppliers without knowing the price it would have to pay until delivery.

It is unclear what actions the sanctions committee took to respond to illicit commissions on commodity contracts. Due to increasing concern about the humanitarian situation in Iraq and pressure to expedite the review process, the Security Council passed resolution 1284 in December 1999 to direct the sanctions committee to accelerate the review process. Under fast-track procedures, the sanctions committee allowed OIP to approve contracts for food, medical supplies, and agricultural equipment (beginning in March 2000), water treatment and sanitation (August 2000), housing (February 2001), and electricity supplies (May 2001).

Issues for Further Investigation

Several investigations into the Oil for Food program are planned or under way. A U.N. inquiry officially began on April 21, 2004, with a Security Council resolution supporting the inquiry[9] and the appointment of three high-level officials to oversee the investigation. This investigation will examine allegations of corruption and misconduct within the United Nations Oil for Food program and its overall management of the humanitarian program. In addition, Iraq's Board of Supreme Audit contracted with the accounting firm Ernst and Young to conduct an investigation of the program. Several U.S. congressional committees have also begun inquiries into U.N. management of the Oil for Food program and U.S. oversight through its role on the sanctions committee.

These investigations of the Oil for Food program provide an opportunity to better quantify the extent of corruption, determine the adequacy of internal controls, and identify ways to improve future humanitarian assistance programs conducted within an economic sanctions framework. Based on our work, we have identified several questions that should be addressed:

- How did the size and structure of the Oil for Food program enable the Iraqi government to obtain illegal revenues through illicit surcharges and commissions?
- What was the role of U.N. member states in monitoring and enforcing the sanctions? What were the criteria used to certify national purchasers of oil and suppliers of commodities?
- What actions, if any, were taken to reduce the smuggling of Iraqi oil? What precluded the sanctions committee from taking action?

---

[9]U.N. Security Council Res. 1538 (April 21, 2004).

- Who assessed the reasonableness of prices for commodity contracts negotiated between the Iraqi government and suppliers and what actions were taken? How were prices for commodities assessed for reasonableness under fast-track procedures?

Much of the information on surcharges on oil sales and illicit commissions on commodity contracts is with the Iraqi ministries in Baghdad and national purchasers and suppliers. We did not have access to this data to verify the various allegations of corruption associated with these transactions. Subsequent investigations of the Oil for Food program should include a statistical sampling of these transactions to more accurately document the extent of corruption and the identities of companies and countries that engaged in illicit transactions. This information would provide a basis for restoring those assets to the Iraqi government.

Subsequent evaluations and audits should also consider an analysis of the lessons learned from the Oil for Food program and how future humanitarian programs of this nature should be structured to ensure that funds are spent on intended beneficiaries and projects. For example, analysts may wish to review the codes of conduct developed for the CPA's Oil for Food coordination center and suppliers. In addition, U.N. specialized agencies implemented the program in the northern governorates while the program in central and southern Iraq was run by the central government in Baghdad. A comparison of these two approaches could provide insight on the extent to which the operations were transparent and the program delivered goods and services to the Iraqi people.

Challenges in Addressing Iraq's Food Security

Evolving policy and implementation decisions on the food distribution system and the worsening security situation have affected the movement of food commodities within Iraq. As a result, warehouse stocks are low, and Iraq has less than a month's supply of several food items, including staple grains, and no buffer stock. The food distribution system created a dependency on food subsidies that disrupted private food markets. The government will have to decide whether to continue, reform, or eliminate the current system. In addition, inadequate oversight and corruption in the Oil for Food program raise concerns about the Iraqi government's ability to manage the food distribution system and absorb donor reconstruction funds under existing structures. The CPA has taken steps, such as appointing inspectors general, to strengthen accountability measures in Iraq's ministries.

Inadequate Planning, Coordination, and Security Have Resulted in Tenuous Food Supplies

The CPA's failed plans to privatize the food ration system and delayed negotiations with WFP on food procurement and distribution resulted in diminished stocks of food commodities and localized shortages in early 2004. The CPA administrator discussed eliminating Iraq's food distribution system and providing recipients with cash payments based on plans submitted to the CPA in summer 2003 that asserted that the system was expensive and depressed the agricultural sector. As a result, the Ministry of Trade began drawing down existing inventories of food. In December 2003, as the security environment worsened, the administrator decided not to reform the ration system. In January 2004, the CPA negotiated a memorandum of understanding (MOU) with WFP and the Ministry of Trade that committed WFP to procuring a 3-month buffer food stock by March 31, 2004, and assuming the delivery of food to hub warehouses inside Iraq through June 2004. The MOU was delayed due to disagreements about emergency food procurement, contract terms, and the terms of WFP's involvement. No additional food was procured during the negotiations, and food stocks diminished and localized shortages occurred in early 2004. WFP completed its buffer stock procurement by March 31, 2004. The Ministry of Trade assumed responsibility for food procurement on April 1, 2004, and will implement the distribution system after June 30, 2004.

A U.S. official stated in early March 2004 that coordination between WFP and the Ministry of Trade had been deteriorating. The Ministry had not provided WFP with complete and timely information on monthly food allocation plans, weekly stock reports, or information on cargo arrivals, as the MOU required. WFP staff reported that the Ministry's data were subject to sudden, large, and unexplained stock adjustments, thereby making it difficult to plan deliveries. A State Department official noted in April 2004 that coordination between WFP and the Ministry was improving. However, according to early June 2004 discussions with other U.S. officials, these coordination problems are continuing.

The security environment in Iraq has affected the movement of Oil for Food goods since the fall of 2003. A September 2003 U.N. report found that the evacuation of U.N. personnel from Baghdad, following the bombing of the U.N. office in August 2003, affected the timetable and procedures for the transfer of the Oil for Food program to the CPA and contributed to delays in prioritizing and renegotiating contracts. The August bombing of the U.N. office also resulted in the temporary suspension of the border inspection process and shipments of humanitarian supplies and equipment. A March 2004 CPA report noted that stability of the food supply would be affected if security conditions worsened.

According to an Oil for Food coordination center official, the worsening security situation during April 2004 affected food supplies. As of early June, major food transport corridors from Jordan and the port of Umm Qasr are restricted due to security concerns, and border crossings from Jordan, Syria, and Turkey are congested. Also, fewer drivers are willing to work in this environment, thereby reducing the movement of food from the borders and ports to the food warehouses. This situation is exacerbated by congestion at the major port of Umm Qasr, which is operating at 50 percent of its capacity due to inadequate fuel and power supply, off-loading delays, dredging activity, inadequate storage capacity, and security concerns.

Initial planning and management problems, combined with security and port congestion issues limiting the movement of food, have resulted in the drawing down of warehouse food stocks. The food supply situation was described as tenuous by several U.S. and WFP sources in early June. At that time, Iraq had less than a 1-month food supply for several items in the food basket, including grains. About 360,000 metric tons of the 1.6 million metric tons procured for the buffer stock had arrived as of June 10, but the full amount will not be delivered until September, according to a WFP official. Moreover, these commodities are not being reserved as a buffer stock, but are immediately used as operating stocks.

U.S. officials are concerned that, as the Iraqi government assumes full responsibility for food distribution on July 1, 2004, it will find it difficult to manage the food distribution system given low food supplies. According to U.S. and WFP officials, the Ministry of Trade implemented the food distribution system during the Oil for Food program under more favorable conditions. For example, the Ministry was able to maintain at least a 6-month food inventory and operate in a more secure environment.

Food Distribution System Essential for Current Food Security But May Not Be Sustainable

The Oil for Food program facilitated the operation of the Public Distribution System run by Iraq's Ministry of Trade. Under this system, each Iraqi is eligible to receive a monthly "food basket" that normally consists of a dozen items.[10]

[10] Wheat flour, rice, vegetable ghee (semifluid clarified butter used for cooking), pulses (edible seeds of various leguminous crops, such as peas, beans, or lentils), sugar, tea, salt, milk, infant formula, weaning cereal, soap, and detergent.

After the CPA transfers responsibility for the food distribution system to the Iraqi provisional government in July 2004, the government will have to decide whether to continue, reform, or eliminate the current system. Documents from the Ministries of Finance and Planning indicate that the annual cost of maintaining the system is as high as $5 billion, or about 25 percent of total government expenditures. In 2005 and 2006, expenditures for food will be almost as much as all expenditures for capital projects. According to a September 2003 joint U.N. and World Bank needs assessment of Iraq,[11] the food subsidy, given out as a monthly ration to the entire population, staved off mass starvation during the time of the sanctions, but disrupted the market for food grains produced locally. The agricultural sector had little incentive to produce crops in the absence of a promising market.

However, the Iraqi government may find it politically difficult to scale back the food distribution system with an estimated 60 percent of the population relying on monthly rations as their primary source of nutrition. WFP is completing a vulnerability assessment that Iraq could use to make future decisions on food security programs and better target food items to those most in need. WFP's preliminary assessment results found that 10 percent of the population was extremely poor and needed food aid in addition to the Public Distribution System. WFP is also developing an emergency operation plan to meet the needs of vulnerable populations. In addition, in April 2004, a USAID contractor submitted a strategy for a short-term plan to stabilize the agricultural sector by providing agricultural supplies, re-establishing domestic wheat markets, rehabilitating irrigation systems, and rehabilitating Ministry of Agriculture facilities. The strategy also includes a medium-term plan to create appropriate agricultural policies, provide capacity building for market-led agriculture, and strengthen the agricultural sector through national programs.

Addressing Corruption

In the absence of significant reforms, the history of inadequate oversight and corruption in the Oil for Food program raises questions about the Iraqi government's ability to manage the import and distribution of food commodities and the billions in international assistance expected to flow into the country. The CPA and Iraqi ministries must address corruption to help ensure that the food distribution system is managed with transparent and accountable controls. Building these internal control and

[11] United Nations/World Bank, *Joint Iraq Needs Assessment: Agriculture, Water Resources, and Food Security* (New York: October 2003).

accountability measures into the operations of Iraqi ministries will also help safeguard the $18.4 billion in fiscal year 2004 U.S. reconstruction funds and $13.8 billion pledged by other countries.

To address these concerns and oversee government operations, the CPA administrator appointed inspectors general for Iraq's 26 national ministries. At the same time, the CPA announced the establishment of two independent agencies to work with the inspectors general—the Commission on Public Integrity and a Board of Supreme Audit. Finally, the United States will spend about $1.63 billion on governance-related activities in Iraq, which will include building an effective financial management system in Iraq's ministries.

The CPA's coordination center continues to provide on-the-job training for ministry staff who will assume responsibility for food contracts after July 2004. Coalition personnel have provided Iraqi staff with guidance on working with suppliers in a fair and open manner and determining when changes to letters of credit are appropriate. In addition, according to center staff, coalition and Iraqi staff signed a code of conduct, which outlined proper job behavior. Among other provisions, the code of conduct prohibited kickbacks and secret commissions from suppliers. The center also developed a code of conduct for suppliers. In addition, the center has begun implementing the steps needed for the transition of full authority to the Iraqi ministries. These steps include transferring contract-related documents, contacting suppliers, and providing authority to amend contracts. In addition, the January 2004 MOU agreement commits WFP to training ministry staff in procurement and transport functions through June 30, 2004. Ten ministry staff are being trained at WFP headquarters in Rome, Italy.

Mr. Chairman and Members of the Committee, this concludes my prepared statement. I will be happy to answer any questions you may have.

Contacts and Acknowledgments

For questions regarding this testimony, please call Joseph Christoff at (202) 512-8979. Other key contributors to this statement were Pamela Briggs, Mark Connelly, Lynn Cothern, Zina Merritt, Tetsuo Miyabara, Valerie Nowak, Stephanie Robinson, Jonathan Rose, Richard Seldin, Audrey Solis, Roger Stoltz, and Phillip Thomas.

Appendix I: Scope and Methodology

We used the following methodology to estimate the former Iraqi regime's illicit revenues from oil smuggling, surcharges on oil, and commissions from commodity contracts from 1997 through 2002:

- To estimate the amount of oil the Iraqi regime smuggled, we used Energy Information Administration (EIA) estimates of Iraqi oil production and subtracted oil sold under the Oil for Food program and domestic consumption. The remaining oil was smuggled through Turkey, the Persian Gulf, Jordan, and Syria (oil smuggling to Syria began late 2000). We estimated the amount of oil to each destination based on information from and discussions with officials of EIA, Cambridge Energy Research Associates, the Middle East Economic Survey, and the private consulting firm Petroleum Finance.

- We used the price of oil sold to estimate the proceeds from smuggled oil. We discounted the price by 9 percent for the difference in quality. We discounted this price by 67 percent for smuggling to Jordan and by 33 percent for smuggling through Turkey, the Persian Gulf, and Syria. According to oil industry experts, this is representative of the prices paid for smuggled oil.

- To estimate the amount Iraq earned from surcharges on oil, we multiplied the barrels of oil sold under the Oil for Food program from 1997 through 2002 by 25 cents per barrel. According to Security Council members, the surcharge varied, but Iraq tried to get as much as 50 cents per barrel. Industry experts also stated the surcharge varied.

- To estimate the commission from commodities, we multiplied Iraq's letters of credit for commodity purchases by 5 percent for 1997 through 1998 and 10 percent for 1999 through 2002. According to Security Council members, the commission varied from 5 percent to 10 percent. This percentage was also confirmed in interviews conducted by U.S. officials with former Iraqi regime ministers of oil, finance, and trade and with Saddam Hussein's presidential advisors.

GAO did not obtain source documents and records from the former regime about its smuggling, surcharges, and commissions. Our estimate of illicit revenues is therefore not a precise accounting number. Areas of uncertainty in our estimate include:

- GAO's estimate of the revenue from smuggled oil is less than the estimates of U.S. intelligence agencies. We used estimates of Iraqi oil production and domestic consumption for our calculations. U.S.

intelligence agencies used other methods to estimate smuggling.

- GAO's estimate of revenue from oil surcharges is based on a surcharge of 25 cents per barrel from 1997 through 2002. However, the average surcharge could be lower. U.N. Security Council members and oil industry sources do not know when the surcharge began or ended or the precise amount of the surcharge. One oil industry expert stated that the surcharge was imposed at the beginning of the program but that the amount varied. Security Council members and the U.S. Treasury Department reported that surcharges ranged from 10 cents to 50 cents per barrel. As a test of reasonableness, GAO compared the price paid for oil under the Oil for Food program with a proxy oil price for the period 1997 through 2002. We found that for the entire period, the price of Iraqi oil was considerably below the proxy price. Oil purchasers would have to pay below market price to have a margin to pay the surcharge.

GAO's estimate of the commission on commodities could be understated. We calculated commissions based on the commodity contracts for the 15 governorates in central and southern Iraq (known as the "59-percent account" because these governorates received this percentage of Oil for Food revenues). We excluded contracts for the three northern governorates (known as the "13-percent account"). However, the former Iraqi regime negotiated the food and medical contracts for the northern governorates, and the Defense Contract Audit Agency found that some of these contracts were potentially overpriced. The Defense Contract Audit Agency also found extra fees of between 10 and 20 percent on some contracts.

Appendix II: Timeline of Major Events Related to Sanctions Against Iraq and the Administration of the Oil for Food Program

Date	Event/Action	Summary
Aug. 2, 1990	U.N. Security Council Resolution 660	Iraqi forces invaded Kuwait. Resolution 660 condemned the invasion and demands immediate withdrawal from Kuwait.
Aug. 6, 1990	U.N. Security Council Resolution 661	Imposed economic sanctions against the Republic of Iraq. The resolution called for member states to prevent all commodity imports from Iraq and exports to Iraq, with the exception of supplies intended strictly for medical purposes and, in humanitarian circumstances, foodstuffs.
Aug. 6, 1990	Operation Desert Shield	President Bush ordered the deployment of thousands of U.S. forces to Saudi Arabia.
Nov. 5, 1990	U.S. legislation	Public Law 101-513, § 586C, prohibited the import of products from Iraq into the United States and the export of U.S. products to Iraq.
Jan. 12, 1991	U.S. legislation	Iraq War Powers Resolution authorized the president to use "all necessary means" to compel Iraq to withdraw military forces from Kuwait.
Jan. 16, 1991	Operation Desert Storm	Operation Desert Storm was launched: coalition operation was targeted to force Iraq to withdraw from Kuwait.
Feb. 28, 1991	Gulf War cease-fire	Iraq announced acceptance of all relevant U.N. Security Council resolutions.
Apr. 3, 1991	U.N. Security Council Resolution 687 (Cease-Fire Resolution)	Mandated that Iraq must respect the sovereignty of Kuwait and declare and destroy all ballistic missiles with a range of more than 150 kilometers as well as all weapons of mass destruction and production facilities.
Jun. 17, 1991	Creation of U.N. Special Commission	The U.N. Special Commission (UNSCOM) was charged with monitoring Iraqi disarmament as mandated by U.N. resolutions and to assist the International Atomic Energy Agency in nuclear monitoring efforts.
Aug. 15, 1991	U.N. Security Council Resolution 706	Proposed the creation of an Oil for Food program and authorized an escrow account to be established by the Secretary General. Iraq rejected the terms of this resolution.
Sep. 19, 1991	U.N. Security Council Resolution 712	Second attempt to create an Oil for Food program. Iraq rejected the terms of this resolution.
Oct. 2, 1992	U.N. Security Council Resolution 778	Authorized transferring money produced by any Iraqi oil transaction on or after August 6, 1990, which had been deposited into the escrow account, to the states or accounts concerned as long as the oil exports took place or until sanctions were lifted.
Apr. 14, 1995	U.N. Security Council Resolution 986	Allowed Iraq to sell $1 billion worth of oil every 90 days. Proceeds were to be used to procure foodstuffs, medicine, and material and supplies for essential civilian needs. Resolution 986 was supplemented by several U.N. resolutions over the next 7 years that extended the Oil for Food program for different periods of time and increased the amount of exported oil and imported humanitarian goods.
Mar. 27, 1996	U.N. Security Council Resolution 1051	Established the export and import monitoring system for Iraq.
May 20, 1996	Government of Iraq and the United Nations	Signed a memorandum of understanding allowing Iraq's export of oil to pay for food, medicine, and essential civilian supplies.
Jun. 17, 1996	United States	Based on information provided by the Multinational Interception Force (MIF), communicated concerns about alleged smuggling of Iraqi petroleum products through Iranian territorial waters in violation of resolution 661 to the Security Council sanctions committee.

Date	Event/Action	Summary
Jul. 9, 1996	U.N. Security Council Sanctions Committee	Committee members asked the United States for more factual information about smuggling allegations, including the final destination and the nationality of the vessels involved.
Aug. 28, 1996	U.S. delegation to the U.N. Security Council Sanctions Committee	Provided briefing on the Iraqi oil smuggling allegations to the sanctions committee.
Dec. 3, 1996	Islamic Republic of Iran Permanent Representative to the United Nations	Acknowledged that some vessels carrying illegal goods and oil to and from Iraq had been using the Iranian flag and territorial waters without authorization and that Iranian authorities had confiscated forged documents and manifests. Representative agreed to provide the results of the investigations to the sanctions committee once they were available.
Dec. 10, 1996	Iraq and the United Nations	Phase I of the Oil for Food program began.
Jun. 4, 1997	U.N. Security Council Resolution 1111	Extended the term of resolution 986 another 180 days (phase II).
Sep. 12, 1997	U.N. Security Council Resolution 1129	Authorized special provision to allow Iraq to sell petroleum in a more favorable time frame.
Oct. 8, 1997	Representatives of the United Kingdom of Great Britain and Northern Ireland to the United Nations	Brought the issue of Iraqi smuggling petroleum products through Iranian territorial waters to the attention of the U.N. Security Council sanctions committee.
Nov. 18, 1997	Coordinator of the Multinational Interception Force (MIF)	Reported to the U.N. Security Council sanctions committee that since February 1997 there had been a dramatic increase in the number of ships smuggling petroleum from Iraq inside Iranian territorial waters.
Dec. 4, 1997	U.N. Security Council Resolution 1143	Extended the Oil for Food program another 180 days (phase III).
Feb. 20, 1998	U.N. Security Council Resolution 1153	Raised Iraq's export ceiling of oil to about $5.3 billion per 6-month phase (phase IV).
Mar. 25, 1998	U.N. Security Council Resolution 1158	Permitted Iraq to export additional oil in the 90 days from March 5, 1998, to compensate for delayed resumption of oil production and reduced oil price.
Jun. 19, 1998	U.N. Security Council Resolution 1175	Authorized Iraq to buy $300 million worth of oil spare parts to reach the export ceiling of about $5.3 billion.
Aug. 14, 1998	U.S. legislation	Public Law 105-235, a joint resolution finding Iraq in unacceptable and material breach of its international obligations.
Oct. 31, 1998	U.S. legislation: Iraq Liberation Act	Public Law 105-338, § 4, authorized the president to provide assistance to Iraqi democratic opposition organizations.
Oct. 31, 1998	Iraqi termination of U.N. Special Commission (UNSCOM) Activity	Iraq announced it would terminate all forms of interaction with UNSCOM and that it would halt all UNSCOM activity inside Iraq.
Nov. 24, 1998	U.N. Security Council Resolution 1210	Renewed the Oil for Food program for 6 months beyond November 26 at the higher levels established by resolution 1153. The resolution included additional oil spare parts (phase V).
Dec. 16, 1998	Operation Desert Fox	Following Iraq's recurrent blocking of U.N. weapons inspectors, President Clinton ordered 4 days of air strikes against military and security targets in Iraq that contribute to Iraq's ability to produce, store, and maintain weapons of mass destruction and potential delivery systems.

Date	Event/Action	Summary
Mar. 3, 1999	President Clinton Report to Congress	President Clinton provided the status of efforts to obtain Iraq's compliance with U.N. Security Council resolutions. He discussed the MIF report of oil smuggling out of Iraq and smuggling of other prohibited items into Iraq.
May 21, 1999	U.N. Security Council Resolution 1242	Renewed the Oil for Food program another 6 months (phase VI).
Oct. 4, 1999	U.N. Security Council Resolution 1266	Permitted Iraq to export an additional amount of $3.04 billion of oil to make up for revenue deficits in phases IV and V.
Nov. 19, 1999	U.N. Security Council Resolution 1275	Extended phase VI of the Oil for Food program for 2 weeks until December 4, 1999.
Dec. 3, 1999	U.N. Security Council Resolution 1280	Extended phase VI of the Oil for Food program for 1 week until December 11, 1999.
Dec. 10, 1999	U.N. Security Council Resolution 1281	Renewed the Oil for Food program another 6 months (phase VII).
Dec. 17, 1999	U.N. Security Council Resolution 1284	Abolished Iraq's export ceiling to purchase civilian goods. Eased restrictions on the flow of civilian goods to Iraq and streamlined the approval process for some oil industry spare parts. Also established the United Nations Monitoring, Verification and Inspection Commission (UNMOVIC).
Mar. 31, 2000	U.N. Security Council Resolution 1293	Increased oil spare parts allocation from $300 million to $600 million under phases VI and VII.
Jun. 8, 2000	U.N. Security Council Resolution 1302	Renewed the Oil for Food program another 180 days until December 5, 2000 (phase VIII).
Dec. 5, 2000	U.N. Security Council Resolution 1330	Extended the Oil for Food program another 180 days (phase IX).
Mar. 8, 2001	Deputy U.S. Representative to the United Nations Remarks to the Security Council	Ambassador Cunningham acknowledged Iraq's illegal re-export of humanitarian supplies, oil smuggling, establishment of front companies, and payment of kickbacks to manipulate and gain from Oil for Food contracts. Also acknowledged that the United States had put holds on hundreds of Oil for Food contracts that posed dual-use concerns.
Mar. 8, 2001	Acting U.S. Representative to the United Nations Remarks to the Security Council	Ambassador Cunningham addressed questions regarding allegations of surcharges on oil and smuggling. Acknowledged that oil industry representatives and other Security Council members provided the United States anecdotal information about Iraqi surcharges on oil sales. Also acknowledged companies claiming they were asked to pay commissions on contracts.
Jun. 1, 2001	U.N. Security Council Resolution 1352	Extended the terms of resolution 1330 (phase IX) another 30 days.
Jul. 3, 2001	U.N. Security Council Resolution 1360	Renewed the Oil for Food program an additional 150 days until November 30, 2001 (phase X).
Nov. 29, 2001	U.N. Security Council Resolution 1382	The resolution stipulated that a new Goods Review List would be adopted and that relevant procedures would be subject to refinement. Renewed the Oil for Food program another 180 days (phase XI).
May 14, 2002	U.N. Security Council Resolution 1409	UNMOVIC reviewed export contracts to ensure that they contain no items on a designated list of dual-use items known as the Goods Review List. The resolution also extended the program another 180 days (phase XII).
Nov. 6, 2002	U.N. Security Council Sanctions Committee	MIF reported that there had been a significant reduction in illegal oil exports from Iraq by sea over the past year but noted oil smuggling was continuing.

Date	Event/Action	Summary
Nov. 25, 2002	U.N. Security Council Resolution 1443	Extended phase XII of the Oil for Food program another 9 days.
Dec. 4, 2002	U.N. Security Council Resolution 1447	Renewed the Oil for Food program another 180 days until June 3, 2003 (phase XIII).
Dec. 30, 2002	U.N. Security Council Resolution 1454	Approved changes to the list of goods subject to review by the sanctions committee.
Mar. 12, 2003	U.N. Security Council Sanctions Committee	Chairman reported on a number of alleged sanctions violations noted by letters from several countries and the media from February to November 2002. Alleged incidents involved Syria, India, Liberia, Jordan, Belarus, Switzerland, Lebanon, Ukraine, and the United Arab Emirates.
Mar. 19, 2003	Operation Iraqi Freedom	Operation Iraqi Freedom is launched. Coalition operation led by the United States initiated hostilities in Iraq.
Mar. 28, 2003	U.N. Security Council Resolution 1472	Adjusted the Oil for Food program and gave the Secretary General authority for 45 days to facilitate the delivery and receipt of goods contracted by the Government of Iraq for the humanitarian needs of its people.
Apr. 16, 2003	U.S. legislation	Public Law 108-11, § 1503, authorized the President to suspend the application of any provision of the Iraq Sanctions Act of 1990.
Apr. 24, 2003	U.N. Security Council Resolution 1476	Extended provisions of resolution 1472 until June 3, 2003.
May 1, 2003	Operation Iraqi Freedom	End of major combat operations and beginning of post-war rebuilding efforts.
May 22, 2003	U.N. Security Council Resolution 1483	Lifted civilian sanctions on Iraq and provided for the end of the Oil for Food program within 6 months, transferring responsibility for the administration of any remaining program activities to the Coalition Provisional Authority (CPA).
Nov. 21, 2003	U.N. Secretary General	Transferred administration of the Oil for Food program to the CPA.
Mar.19, 2004	U.N. Secretary General	Responded to allegations of fraud by U.N. officials that were involved in the administration of the Oil for Food program.
Mar. 25, 2004	U.N. Secretary General	Proposed that a special investigation be conducted by an independent panel.
April 21, 2004	U.N. Security Council Resolution 1538	Supported the appointment of the independent high-level inquiry and called upon the CPA, Iraq, and member states to cooperated fully with the inquiry.

ANSWERS TO SUBMITTED QUESTIONS

When describing the shortage of buffer food stocks in Iraq in your testimony and during the hearing on June 16, you describe the security situation as one of the main reasons. Is the problem with the delivery of food to warehouses occurring in all parts of Iraq? Are there regions in Iraq that are better off than others?

According to USAID and WFP officials, information on food supply and buffer stocks has been an ongoing concern for those working on the Public Distribution System. They have never been totally confident that the supply numbers provided by the Iraqi government were accurate and, therefore, continue to be unsure as to exactly how much food is in any one place at any one time. The government has often shuttled food from one spot to another to cover shortages, further obscuring the overall food supply picture. Recently, threats of shortages were compounded by a reduction of deliveries from ports of entry to the warehouses, mostly along the major southern and western corridors. They report that, as of the end of June, the situation had improved and more food was moving than had been the case during May.

Regarding which regions may have a better food supply, it is difficult to say with precision given the lack of accurate data. However, officials believe that the northern three Kurdish governorates have tended to have fewer problems than central and southern Iraq. The areas around Fallujah and Ramadi currently are more likely to have supply problems than the rest of the country, given the uncertain security environment.

Can you update the committee on the progress that has been made with regard to increasing the buffer stocks since the hearing on June 16?

According to USDA officials working with the ministries in Baghdad, as of July 7, 2004, Iraq continued to have a month's supply of imported wheat. They reported that the domestic wheat harvest was under way and will soon add a 2-month supply of domestic wheat to existing stocks. WFP-procured food appears to be moving without major difficulties at this time, but it is unclear whether this food will be reserved as buffer stock or immediately delivered to fill ongoing Public Distribution System needs. Monitoring current food supplies is hampered by the current security situation.

Please provide a list of participants (and agency affiliation) involved in the interagency review process related to the "Oil for Food" program.

The primary U.S. Government entities that were involved in the interagency review process are the Departments of

- Treasury (Office of Foreign Assets Control),
- Commerce (Bureau of Export Administration),
- State (Bureaus of International Organization Affairs, Non-proliferation, Near Eastern Affairs, Economic and Business Affairs and U.S. Mission to the United Nations),
- Defense (Threat Reduction Agency and Navy),
- Energy,
- and Justice.

The intelligence agencies were also involved in the process. We do not have the names of individuals.

Oil purchases within Jordan received greater discounts on Iraqi oil compared to other countries. Specifically, Jordan received a 67 percent discount, and the Persian Gulf and Syria received a 33 percent discount. Why does this disparity exist, and what incentive might the former Iraqi regime have had to provide such discounts? Does a pattern exist between third party purchasers of oil who used Jordan vs. other nations?

According to oil experts, 33 percent was the typical discount for Iraqi smuggled oil, but the oil smuggled to Jordan was more heavily discounted. The discount helped Iraq maintain trade with Jordan and do business with Jordanian financial institutions. Jordan stated in a letter to the U.N. Security Council that the sanctions severely hurt its economy and that it depended on favorable trade with Iraq. Because the Oil for Food program controlled the price of oil, Iraq had to smuggle oil to Jordan to allow the discounted price. The sale of Iraqi oil to Jordan was further established in a trade protocol between Jordan and Iraq outside of the Oil for Food program. The U.N. Security Council noted the protocol, but took no further action. We have no information on a pattern between third party purchasers of oil through Jordan vs. other nations.

According to the testimony of Mr. Christoff, the GAO had access to renegotiated Oil for Food contracts. Please submit copies of those contracts that the GAO has reviewed.

GAO did not have access to any of the actual contracts, amendments, or letters of credit associated with any Oil for Food contracts. GAO reviewed and reported on summary information provided by the U.N. Office of the Iraq Program (OIP) on the 3,059 contracts—including 2,808 renegotiated contracts—turned over to the Coalition Provisional Authority (CPA) on November 21, 2003. According to several CPA and U.S. government officials, the CPA did not receive a complete set of original contracts, amendments, or letters of credit from OIP for these contracts. The CPA also stated that the disks containing the database of contract documents provided by OIP contained errors and inaccessible files. A senior OIP official, however, stated that his office submitted all required information, including an operational database, to the CPA at the time of the program's transfer. The CPA provided GAO with copies of the disks it received from OIP. These disks contain thousands of files with contract-related information. However, the disks are not organized to facilitate analysis, and many files are inaccessible.

The DOD Inspector General's Office and the Iraqi Board of Supreme Audit, through a contract with Ernst and Young, are in the process of securing Oil for Food documents in Iraq. The Committee should consider requesting copies of renegotiated contracts from these entities.

The United States was a member of the sanctions committee, which was responsible for reviewing commodity contracts, and a U.S. interagency team reviewed these contracts, as noted above. The Committee should consider requesting copies of these contracts from the State Department.

The State Department provided GAO a list of the approximately 30,000 commodity contracts for all 13 phases of the Oil for Food program. The list is in hard copy and includes the contract number, country, company, the type of goods, and the contract amount. State informed GAO that it would consider making the list available to others upon request. Accordingly, the Committee should consider requesting a copy of this list from the State Department.

Testimony of the USA Rice Federation
And the U S Rice Producers Association

By
John King, III
Vice Chairman of USA Rice Council
Before
The Committee on Agriculture
U.S House of Representatives
Review of Iraqi Agriculture: From Oil for Food
To the Future of Iraqi Production
Agriculture and Trade
June 16, 2004

Introduction

Mr. Chairman and members of the Committee, my name is John King. I am a rice and soybean farmer from Helena, Arkansas. I am testifying today on behalf of both the USA Rice Federation and the U S Rice Producers Association. The USA Rice Federation's Charter Members include the USA Rice Council, USA Rice Producers' Group, and the USA Rice Millers' Association. I currently serve as Vice Chairman of the USA Rice Council and I am a member of the Board of Directors of both the USA Rice Producers' Group and the USA Rice Federation.

I am pleased to appear before the Committee today on behalf of the rice industry. On a personal note, I will share with you how excited and honored I am to testify before Congress for the first time.

Background

Rice production and marketing is a multi-billion dollar activity in the United States. Rice is produced on over 3 million acres in the U.S. and accounts for $1.4 billion in farm revenues. The U.S. rice industry exports 50 percent of its crop to markets around the world. The economic health of U.S. rice farmers and the rice-milling sector is tied to maintaining and expanding access and sales in foreign markets.

The U.S. rice industry wants to play a major role once again in supplying rice to Iraq. With the current challenges facing the U.S. rice industry, particularly in the milling sector, renewed Iraq market access could have a tremendous impact in value added sales and enhanced economies in rural America.

In the late 1980s, Iraq was the number one market for U.S. rice. The United States provided nearly 90% of Iraq's rice imports, which totaled over 500,000 metric tons in peak years. In 1991, the U.S. government imposed export sanctions and this market was lost. Rice is no stranger to this situation as Cuba and Iran were also once top export markets ultimately lost to government-imposed sanctions. The impact of losing our number one export market had a significant negative impact on the U.S rice industry and a positive impact on the rice industries of major competing rice export countries, primarily Vietnam and Thailand.

U.S. Embargo on Iraq Adversely Impacts U.S. Rice Exports

From 1980 to 1990 the United States accounted for about 80% of Iraq's total rice imports. As a result of the embargo on exports to Iraq, and based on the U.S. share of Iraq's rice market in the 1980s, we estimate that the United States lost about $1.9 billion in rice export sales to Iraq from 1991-2003 (Table 1).

Table 1. Estimated U.S. Rice Export Losses to Iraq from Embargo (1,000 MTs)

Year	1,000 Metric Tons	$Million
1991	215	76
1992	439	161
1993	518	176
1994	51	19
1995	79	30
1996	187	71
1997	596	220
1998	504	170
1999	624	213
2000	1,020	270
2001	768	184
2002	943	191
2003	538	124
1991-2003 Total	**6,482**	**1,905**
1991-2003 Avg.	**499**	**147**

Note: Average export prices used in calculation from FAO & U.S. Bureau of Census export data.

From 1991-2003, U.S. rice competitors dramatically expanded sales to Iraq (Chart 1). Thailand, China, and Vietnam are now important exporters to Iraq, and the Iraq market has helped support production in the competitors of the United States.

Chart 1. Iraq Rice Imports from the United States and Other Origins
1980-2003

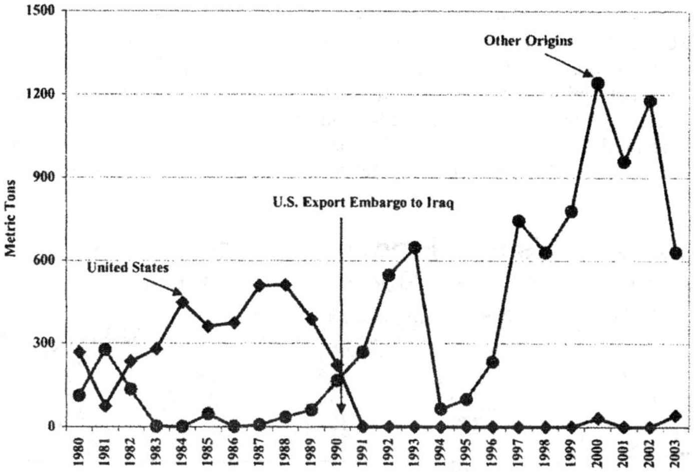

Current Market Conditions

The liberation of Iraq in 2003 by Coalition forces has brought freedom to the Iraqi people. The resumption of trade has also provided hope to the U.S. rice industry. We are anxious to restore our trading ties with the people of Iraq. The Iraq market has grown from about 550,000 metric tons in 1991 to about 1.1 million metric tons in 2004. USDA estimates that Iraq will need to import over 1 million tons of rice per year for several years.

In the summer of 2003, following the war, USDA shipped over 40,000 metric tons of U.S. rice via food aid to the people of Iraq. This was the first significant shipment of U.S. rice to the region in over 14 years. Unfortunately, it remains the only U.S. rice that has entered the region since the liberation. We understand that there are no plans for major food aid programs into Iraq. However, given the potential unstable environment facing Iraq during the upcoming transition of power, we stand ready to supply rice to any targeted food aid efforts that may be required to help vulnerable groups.

While we would welcome the opportunity to assist in providing rice as part of U.S. food aid to Iraq, we are more interested in re-establishing long-term commercial trade between the U.S. rice industry and Iraq.

Oil for Food

The Oil for Food Programme, administered by the United Nations, was the main source of Iraq's food supply from 1991-2003 following the first Gulf War. A small amount of U.S. Rice was shipped to Iraq under this program in 2000. Rice supplied for Iraq under the Oil for Food Programmed was exhausted in the spring of 2004 after the Oil for Food Programme was terminated. The U.S. rice industry was hopeful that the Coalition Provisional Authority would turn to the United States or other coalition partners for its food supply when commercial tenders resumed. Unfortunately this did not occur as the first non-Oil for Food tender of 70,000 metric tons was granted to Vietnam in March of 2004. Subsequent tenders were awarded to Thailand totaling over 300,000 metric tons.

Table 2. Iraq's Rice Production, Imports, and Consumption, 1980-2004 by 5-year averages

(All figures are in 1,000 metric tons and on a milled basis)

	1980-84 avg	1985-89 avg	1990-94 avg	1995-99 avg	2000-04 est
Production	98	115	158	175	85
Imports	370	454	325	732	1,002
Imports from U.S.	**279**	**401**	**0**	**6**	**8**
Consumption	464	554	503	828	1,166
Import share of consump.	80%	82%	65%	88%	86%
U.S. share of imports	**75%**	**88%**	**0%**	**1%**	**1%**
Primary Supplier	U.S.	U.S.	Vietnam, Thailand	Vietnam, Thailand	Vietnam, Thailand

Market Access Development

Re-establishing a viable commercial trading relationship between the U.S. and Iraq must be the goal of rice producers, processors, exporters, and the federal government. To help facilitate this, leaders from the U.S. rice industry traveled to Amman, Jordan, in February 2004, to meet with members of the Iraqi Grain Board in an effort to restore a trading relationship between the two countries. During his opening comments at the start of the summit, Iraq's Minister of Trade Director Ahmad Al-Mukhtar said the Iraqi people have good memories of past purchases of high quality U.S. rice. The summit attendees discussed technical information about grades and trade specifications, delivery time periods and delivery destinations preferred by Iraq. A follow up meeting between Iraq officials and the U.S. rice industry was recently postponed, but we are hopeful that it will be rescheduled soon.

Market development will be vitally important in the months and years to come as U.S. agriculture attempts to re-establish its markets in Iraq. For this reason, we urge this Committee to stress the importance of the Market Access Program (MAP) and Foreign Market Development (FMD) Program to the appropriators as the spending bills are prepared for fiscal year 2005. It is imperative that Congress provides full funding for MAP and FMD at levels authorized in the 2002 Farm Bill, if sufficient funds are to be available to meet current needs and re-establish our markets in Iraq.

It is critical that GSM export credits be made available for exports to Iraq as soon as possible. Some in the Administration have suggested that no new GSM-102 financing can be provided to Iraq since the former regime is in arrears. However, it is our understanding that under current law having an outstanding debt or a rescheduling arrangement does not preclude USDA from offering a country new GSM-102 credit. There is no statutory impediment to prevent the Department from offering a GSM credit program to Iraq. Given this situation, we strongly recommend that the USDA announce its intention to establish a program for GSM-102 credit for Iraq as soon as possible. This credit package is a critical component to helping the U.S. rice industry rebuild its presence in this major market.

We are sympathetic to the need to address Iraq's outstanding debt to the United States, and we are committed to providing whatever assistance we can to accomplish this goal. But, we are of the strong opinion that Iraq's debt issues can and should be addressed concurrent with the provision of new GSM credits. The extension of credit does not need to, and should not be postponed until the existing debt issues are addressed. The future prosperity of U.S. agriculture should not be jeopardized by debts piled up by the Saddam Hussein regime.

Conclusion

In summary Mr. Chairman, Iraq was a top export market for the U.S. rice industry in the past, and we hope it will be again. We urge Congress to support market development programs to help us rebuild relationships in Iraq, and we seek GSM credit assistance to facilitate future sales.

Thank you again for the opportunity to testify here today. I would be pleased to address any questions that you may have.

STATEMENT OF LOCHIEL EDWARDS

Good Morning Chairman Goodlatte and members of the committee. My name is Lochiel Edwards and I farm on the prairies of Montana. My sister, brother, and children work together with myself to raise high quality wheat for domestic flour mills and for overseas export. I am currently President of the Montana Association of Wheat Growers.

Today I am representing the Wheat Export Trade Education Committee, the trade policy arm of the wheat industry; the National Association of Wheat Growers, the organization responsible for domestic policy and farm programs; and for U.S. Wheat Associates, the industry's foreign market development and promotion organization. I appreciate this opportunity to speak to you on where we now stand and what needs to be done to rebuild the Iraq wheat market.

Let me begin by highlighting two points that wheat producers in the United States take into account when looking at export opportunities around the world. First, 96 percent of the world's consumers live beyond our border. The four percent within the United States do not consume enough wheat to sustain a viable wheat industry.

Second, we consistently export nearly 50 percent of our total production. As you can imagine, our success or failure hinges on the ability of U.S. wheat to be exported around the world. Trade is a vital component for ensuring the financial viability of U.S. wheat farmers. Every market, regardless of size, is an important market.

As early as 1963, Iraq was an important customer of the United States wheat industry. There was a year in the 1970's when we had 100 percent market share. In the mid and late 1990's there were some exports to Iraq. However, in most years leading up to the 1991 Gulf War, we maintained about a third of the total annual Iraqi wheat import market at around 3 million tons per year. During the first Gulf War and the period following Iraq refused to make any purchase from the U.S. not only because of the war, but also, because of the unilateral sanctions we maintained against imports of their products. Regaining market share is difficult at any time. The challenges in a war torn country will no doubt add to the difficulties of regaining market share that has been displaced by our competitors.

We are very pleased that a new market-oriented Grain Board is up and running. Even with renewed ties from the past, it will be tough going for U.S. wheat to regain market share. We are going to need the assistance of the U.S. Government.

I'd like to take a couple of minutes to explain some of the wheat industry's relatively recent history in Iraq, and then look to the immediate and long-term future.

During the 1980's, through 8 long years of war between Iraq and Iran, Iraq was a high priority market for our industry and U.S. Wheat Associates' put a great deal of work into servicing this market. Dealing with the U.S. free market system was more difficult for Iraq than the government-to-government system they were able to use with the Canadian Wheat Board and the Australian Wheat Board.

While not able to sign long term purchase agreements with Iraq like our competitors could offer, the U.S. wheat industry signed a protocol with Iraq's State Organization for Grains (SOG), encompassing technical cooperation. The agreement provided for an in-depth comprehensive program of technical assistance, trade servicing and information, in return for SOG's agreement to give the U.S. wheat market consideration as a major source of supply in meeting Iraq's annual imports.

With the support and backing of the U.S. Department of Agriculture, we were able to offer several months of training to key officials of the SOG. Iraqi millers were trained in Kansas and many bakers were awarded scholarships to 19-week courses at the American Institute of Baking. They received assistance that allowed them to computerize their grain inventory system. Key Iraqi grain inspection officials received training at the North Dakota's Northern Crops Institute and key grain purchasers received training, with the financial support of USDA, in grain marketing.

By 1989/90, allocation of our market development funds for Iraq was three times more than funding for activities in Egypt, the region's largest wheat importer. This was done because the potential market in Iraq was growing quickly.

Unfortunately, with the escalation of the Iraq-Iran war Iraq's industry representatives were unable to attend any activities outside of Iraq. To meet this constraint between1988 and 1990, U.S. Wheat Associates offered training in grain inspection, transportation, storage, milling and baking within the country. In 1990/91, political strains and eventual war halted all U.S. market development efforts.

Some U.S. hard red winter wheat went to Iraq for a couple of years in the mid 1990's, until May 1998, when Iraq's regime said it would refuse to grant contracts to companies in America. That was an abrupt end to a wheat marketing year that saw exports of 817,400 metric tons of U.S. wheat to Iraq.

With an opening of the market following the end of the Hussein regime, the U.S. wheat industry is working to renew friendships and pick up where we left off in offering all the assistance they need. We weren't and are not looking for special deals. In fact, we believe that any allegations of improper past arrangements should be fully investigated.

We strongly believe that Iraqis and Americans alike benefit from the development of an open and competitive marketplace where economics and specific end use needs rather than political considerations determine what type of wheat is bought and from where. Open markets work, when given the chance. They work best when commercial buyers are free to choose their sources of supply and when suppliers compete fairly and transparently.

Unfortunately, we've faced many roadblocks on the road to normal trade relations. For nearly a year, despite the fact that our competitors were allowed into Iraq, our government prohibited us from travel into the Iraq. Finally, in February of 2004, almost a year after the war began, our market development experts were able to meet with Iraqi wheat buyers. The process of rebuilding friendships and providing assistance has begun.

The future for this market is positive. The Iraqis are enthusiastic about once again working with us. U.S. Wheat Associates was encouraged by a meeting with Iraq's Grain Board, Foodstuffs Trading Board and the Ministry of Trade that was held in Jordan this year. They clearly wanted to learn as much as they could about getting back into the U.S. marketing system. Our team in Iraq believes that they want a competitive purchasing system, and they want to have the U.S. as a supplier and competitor.

Iraq buyers have made a lot of progress in understanding wheat quality specifications, and we believe that it will only get better as the Iraqis become reacquainted with the U.S. marketing system and the role of the Federal Grain Inspection Service in providing for official certification of quality at loading. There have been roadblocks in recent attempts to bring Iraqi teams to the U.S. for training. One of these has been difficulty in getting visas issued in a timely fashion, even when the individuals were invited and sponsored by the U.S. government, and representatives from the government were to accompany the team. Efforts were made to bring a team to the U.S. in April for training, but the clearances could not be obtained. Belatedly, the visas were issued, and we are now trying to reschedule the trade mission.

Another problem is the lack of ability to extend USDA's credit program. Iraq's debt will need to be restructured so that the country can begin to recover without the overhang of an estimated $150 billion. As I stated in the beginning of this testimony, it will take a lot of effort to re-establish U.S. wheat in this previously closed market. Our industry will do all that we can. But we need the assistance of the U.S. Government in the following areas:

(1)Work with the other countries in debt forgiveness and debt restructering;

(2) Allocation of GSM credits;

(3) Expeditious approval of visas for approved Iraqis to visit the United States;

(4)USG funding for market development activities and improvement in Iraq's port and grain handling facilities;

(5) Travel authority to visit Iraq for market development activities;

(6) The UN and other bodies are investigating alleged kickbacks of the previous Iraq regime. The U.S. Government needs to firmly support that effort.

I am very pleased to be able to report that Iraq has purchased U.S. wheat this year. As the Iraqis regain their familiarity with the U.S. marketing system, and rediscover the unique qualities of U.S. wheat, we are confident of a brighter future. We look forward to a time in the near future when we once again control a strong share of this important market.

○